Backspring

Backspring

Judith McCormack

A JOHN METCALF BOOK

BIBLIOASIS
WINDSOR, ONTARIO

FIRST EDITION

Library and Archives Canada Cataloguing in Publication

McCormack, Judith, 1954-, author
Backspring / Judith McCormack.

Issued in print and electronic formats.
ISBN 978-1-927428-87-0 (pbk.).--ISBN 978-1-927428-88-7 (ebook)

I. Title.

PS8575.C668B33 2015 C813'.6 C2014-907961-3
C2014-907962-1

Edited by John Metcalf
Copy-edited by Allana Amlin
Typeset by Chris Andrechek
Cover designed by Kate Hargreaves
Cover image by Juli Puli (behance.net/julipuli)

Canada Council for the Arts
Conseil des Arts du Canada

Canadian Heritage
Patrimoine canadien

Published with the generous assistance of the Canada Council for the Arts and the Ontario Arts Council. Biblioasis also acknowledges the support of the Government of Canada through the Canada Book Fund and the Government of Ontario through the Ontario Book Publishing Tax Credit.

PRINTED AND BOUND IN CANADA

For Dayal Kaur Khalsa, Michelle Swenarchuk and Dianne Martin—three friends who died too early.

Chapter One

Standing alone on the edge of the white rock, exposed in the brilliant light of the Algarve—head-catching light, bright as a fact, bouncing off the saltwater. The breeze slips around him, across his wet skin, raising goose bumps. His skinny arms are pulled in tight to his body, fists clasped together in front of his chest—the involuntary prayer of the shiverer. He can hear the shouts of the other boys echoing off the pale cliffs that curve around this shoreline. How many times has he jumped off this rock? Ten—fifteen?—times this morning alone, kicking his legs in the air, then plunging down into the soundless green of the water and wriggling his way up to the surface again. But now—suddenly—his feet are clamped to the rough rock, and the distance to the water below seems unimaginably long. His head is buzzing with vertigo—an attack of it so swift and unfamiliar that he thinks it must be fear.

The shouts of the other boys are closer now, they are scrambling up the rock to see what he is looking at—a sand snail? Sea asters? A dead snake? He tries to unstick his feet, to push himself off, but his legs feel hollow and soft.

Then the boys are crowding around him, as inquisitive as sandpipers. When they see nothing there but his shadow in the sun, they go back to jumping off the rock themselves, whooping as they go. An older boy eyes him and his rigid posture, then quietly moves over beside him and pats his back kindly, murmuring something. Then the pat turns into a shove and he pushes the panicked boy over the edge of the rock, out into the air, falling, falling—until he crashes through the broken edges of sleep, and into a morning of city heat.

He lies there on the bed for a few seconds, trying to unclench his muscles, to slow his pounding heart. Eduardo Aguiar Cabral, a large man, subtle as hydrogen, curious as a question mark—an unwilling contestant in his own life. Someone with a taste for catastrophe, although this particular love has so far been unrequited.

He has woken up in the middle of a thought. What thought? Who knows? One of those vague impressions, a shapeless sense of something unfinished. A thought that skitters away before he can remember it, leaving him feeling uneasy.

He reaches out a hand without looking and gropes the other half of the bed. No, she has left already, although he can smell her clove-scented soap. Or maybe the smell is coming from the small clay bottles of clove oil she puts in their closets—to ward off moths, she says. Is that why she uses the soap on herself? To deter unwelcome intruders? It wouldn't surprise him.

He stretches his legs out onto her side of the bed, a rare luxury for a tall man. She must be lecturing this morning—she was roped into teaching a summer course this year. Botany—to students, who are, if anything, too interested, she says. Or too metaphysical or something.

They come in thinking that cell composition holds the answers to everything, that some strand of DNA will show them the meaning of their existence. Where do they get this from? They all seem to believe in it. I think it's contagious. I think it's a kind of virus. They should be taking virology.

She says this a little wearily, though, as if she finds their expectations almost physically demanding. Repeat after me. A rose is a rose is a rose. A leaf is a leaf is a leaf.

But she doesn't really believe this herself, he thinks. No, she wants her students to be captivated by the artistry of cell structure in itself, by peroxisomes and ribosomes and nuclei. She doesn't like the idea of these things being stand-ins for some watered-down philosophy. Well, she specializes in fungi, anyway.

He is suddenly impatient with this Geneviève-less bed, and besides, if she is gone, then he is late.

Get up, get up, get *up*—he rolls his heavy body out of bed. He is a solid man, but not fat—his heaviness has more to do with physical density than anything else.

Over the washbasin, he hesitates, running his hand over his beard stubble, examining the creases in his face, his faded olive skin. Then he picks up the razor and shaves quickly—too quickly—barely avoiding a nick. Clothes? The grey silk, the lightest suit he has. Still too warm for a summer day like this, but necessary. Can he skip the jacket? No, not for a first meeting, not for a potential client-to-be. If clothes don't make the man, they can at least give him a hand, and he needs all the help he can get. (There's something formal about you anyway, says Geneviève. Suits suit you.)

She has left him some coffee, but the pot is cold—he can get something at the market. He grabs a croissant and then he is out of the apartment, down the stairs, too impatient for the elevator. The car is a block away, and the air inside it is already warm. When he turns the key, it hesitates before the engine catches. Christ, something else he has been putting off, something else overdue. He can drop it at the garage on the way.

By the time he reaches the first intersection, his sweat glands are starting to prickle.

'*August*,' he says, as if it were a profanity. A month he detests, a month that seems particularly unreliable. Absurd? Maybe, but isn't he entitled to a few absurdities? A man who at least tries to be rational—whatever that means—for most of his waking hours should be allowed some leeway, a stupidity or two.

And August in Montréal—if anything, particularly unreliable. Why? This is a place that has several different summers at the same time—some spread out across the city, changing from block to block, others overlapping, climbing on top of each other. Maybe the better word is *unpredictable*. Unpredictable, unreliable—whatever it is, the result is this kind of morning, tense from the start.

It doesn't help that the driver in front of him keeps swerving unexpectedly, as if he considered driving a form of suicidal entertainment. The cars behind him begin honking, and he leans on his own horn. The traffic gladiator. May the loudest horn win.

In a few minutes he pulls into the garage lot, his stomach already tight. His father comes out to meet him, wiping his hands on a greasy rag.

'*Olá*.'

The older man seems pleased in his leathery way, his short hair—greying to white—standing up in tufts. (He looks as if he's been salted and cured, said Geneviève the first time she met him.)

He claps Eduardo on the back, then grabs the nape of his neck, shaking it a couple of times. He has to reach up to do this—Eduardo is almost a head taller. Not that his father minds—irritatingly, he seems to consider his son's height his own accomplishment. Almost a remedy for one of the indignities of immigration—he left Portugal as a standard-sized man and arrived here to find he was short. A height disadvantage acquired in the space of a plane ride. But Eduardo—Eduardo is proof that he has bested this ridiculous country at its own game.

Why begrudge him this? thinks Eduardo. But he does.

He watches his father spring the catch on the car hood, and begin poking around in the engine, his hands moving in a slow, confident rhythm. For a second, Eduardo almost longs to be back here—the dim interiors of the garage bays, the clanging and hissing of tools, the laconic scraps of conversation. At least there is a kind of grimy logic to things here, a way of making sense. The problems of leaking tires, spongy brakes, dead batteries are usually knowable, usually fixable. Patch it, grease it, weld it, replace it.

Reassuring for a nine-year-old, this string of questions and answers—particularly a nine-year-old stunned with dislocation. *What's the problem? Here's the problem. What's the solution? Here's the solution.* And something to distract him from the humiliation of his stumbling English, the shirts and trousers like his father's, the incriminating smell of his *linguiça* sandwiches at school.

He would sit on a workbench while his father ran him through the four stroke engine cycle. One—the intake of air and gas. Two—the compression in the cylinder. Three—the ignition of the mixture and four—the expulsion of exhaust. Look, his father would say lovingly—*suck, squeeze, bang, blow*, a formula he had picked up from the

English mechanics, with no idea how vulgar it was. But then Eduardo had abandoned the garage for other interests—still a sore point, this many years later.

His father is talking now, bending over the engine in the hot sun. He points out things with two thick fingers, his other hand on the edge of the car, still holding the rag. Behind him in the garage, a pneumatic wrench hisses in short bursts.

'Fine, fine,' says Eduardo. 'I have to go. I'll pick it up later.'

'*Vá embora,*' says his father, disgusted. Go. What kind of man is not interested in his own car?

Eduardo is instantly annoyed in turn, and then annoyed with himself. Surely a man in his forties should know better than to react to his father? Yes, he should. Does he? No.

He sets off across the street, walking quickly towards the market. A man in his forties—how did this decade turn out to be so tricky anyway? He is still startled to find that he has arrived here, as if he had put his thirties down absently, and forgotten where they were. Where did they wander off to? Will they return at some point, sheepish and apologetic? No, he has had his chance with them.

Don't be so bleak, says Geneviève. Getting older is only fatal at the end.

You can afford to make fun of it, he says. She is younger than him.

You can't afford not to, she says, laughing.

Geneviève—small-breasted, thin-armed, her face soft and irregular, her dark hair thick but so fine it sticks to her clothes. A sensitive spine, straight shoulders, a brown birthmark on her elbow. He feels his groin begin to stir at the thought of the smoothness of her hip bones, the tendons on the inside of her thighs. She has an odd radiance that comes from believing she is beautiful, a belief that is almost convincing in itself. An exuberant streak, too, something it took him a while to understand. Now he knows where it comes from—so many things strike her as unaccountably hilarious.

But he should be thinking about this meeting, coming up with something for the client—or strictly speaking, the client-to-be. He checks his watch—less than an hour to go. He needs something to say to him, something that sounds like he has a design in mind, or at least some ideas. What will hook this man, what will grab his imagination and hold it? It would help if he knew something about him—the telephone call was not very illuminating. A high, nasal voice, accented English—eastern European? *I've seen your other work.* A quick reference to the proposed project—renovating an old market building he owns, the grounds around it. Solve the problems of the aging building, and make it more of a destination, instead of simply a place where people pick up their groceries. (What's wrong with a place where people pick up their groceries? says Geneviève. A good, honest purpose.) Then they made the arrangements for the meeting at the market, in a tiny café on the second floor. Of course, he will find out more at the meeting, but he needs something to go in with, something to wave around in front of the man's nose.

The air is already tight and sticky, and he stops to loosen his tie.

Ideas—not usually his problem. If anything, he suffers from a surplus. But this thing has been resisting his efforts—he keeps coming up dry. His usual method is to herd his thoughts into a specific area of his brain, and then hope they will combine into new sequences, new configurations. He knows a certain incubation period is required for this, but normally he would have something by now—vague, maybe, but something. Not here, though—the project is still shimmering off in the distance, out of reach in a peculiarly frustrating way.

For lack of anything better, he has been collecting material about other street markets—floor plans, photographs of the Naschmarkt in Vienna, the Cours Saleya in Nice, the Balik Pazari in Istanbul, more. Design fodder. He is starting to feel he is on an intimate basis with these markets—he has tracked their inner workings, their technical undersides. But despite this, he hasn't been able to come up with even a starting point for a discussion, even a wisp of an idea.

He is not only stumped by this, he is stumped as to why he is stumped. Could street markets be somehow resistant to planning? Maybe they are naturally haphazard—most of them seem to have grown up piecemeal on the banks of muddy rivers, spreading out along wharfs, creeping through alleys. If he closes his eyes, he can see a table of black plums, the vendor holding one up and spinning it around in his hand. A woman waving a bee off a slab of pork. A boy sluicing water on the pavement around buckets of Parma violets.

But he isn't about to concede that anything could be truly resistant to design. What architect believes that? Certainly not him, someone for whom design is almost an article of faith, a way of being—something that inhabits everything else, subsumes everything else.

Presumptuous? Maybe. Deluded? Probably. He doesn't care.

As he walks along, a picture of the street in front of him is automatically reeled up into his brain, to be stored for future reference. Do other architects do this? Perhaps, although he has never asked anyone. What if they said *no, we don't do that*? Or worse, if they said, *yes, we do that, too.*

Reeling up at the moment are limestone houses, three stories with outdoor staircases twisting down in front of them. A bicycle locked to a balcony, a rental sign pasted up in a window—*appartement à louer.* Elderly trees hang over the sidewalk and a weak breeze ruffles their leaves for a few seconds, showing their silvery undersides. The grey of the limestone, the grey of the sidewalks make this part of the street seem austere.

Walking was a good idea, though—something about this seems to channel his uneasiness, turn it into momentum. He is restless, anyway, even at the best of times—harassed by a need for newness, a yearning for the unusual, the unexpected. A yearning? Almost a compulsion, even an addiction of sorts—that same soft, slinky hunger, then the sense of gratification. Although he assumes most people have this to some degree—he simply has an acute case.

How acute? Enough that he catches himself occasionally making almost unconscious calculations of what he has, a tally of his firm, Geneviève, friends, all the moving parts of his current life—and what he might lose if he suddenly veered off course.

He would be hard put to say what produces this restlessness, or even—if this is an addiction of sorts—what it is he craves. Easier to describe things that get to him: a crack of heat lightning, the shock of ice water, even a new kind of silence, a collection of soundless molecules. Anything to feed his novelty-starved mind, to hit that group of brain cells that produces relief from boredom. He isn't interested in what passes for happiness, something that seems more like a smog of contentment. He wants something wilder, more startling.

Or perhaps he wants both.

A corner store is coming up quickly on his left, *Dépanneur Gauthier* across the top, the windows full of faded signs: *Bière froide, cidre, vins. Loteries. Crème glacée.* Behind them, he can see shelves filled with cases of beer, cigarettes, rows of manufactured food—salty, fatty, sweet. All the cheap things that soothe or satisfy people, from the first cold tongue-full of ice cream to the last pale chip, glistening with oil.

Where is this market, anyway?

He wants to get there before the meeting to give himself time to wander around first, to immerse himself in the sights and sounds of the market. Maybe this will help, maybe it will unlock something in his brain. Maybe it will at least give him something to say.

But why does his brain need unlocking? And now of all times—when the firm really needs the work. Well, the firm always needs work. But now when the firm needs work so badly.

He pushes away the thoughts that are starting to hang around him faintly, like a watermark in the air. (What if this dry spell isn't temporary? What if he has lost his touch, his ability to conjure up designs?) No, impossible. If his thoughts are stuck at the moment, the moment will pass.

A gull screeches overhead. He must be getting close.

He starts trying to assemble some phrases for the owner. What about combining the lively jumble of a market with a certain risky elegance? What about considering the crowd of people as a kind of living entity—something that the right surroundings could coax along, or question, or court? Why not make the most of the fact that the market has a physical flow to it, a place that exists inside and outside at the same time?

Architect-babble, says Patrick.

As if lawyers don't babble, says Eduardo.

Notaries, says Patrick automatically. *Caveat emptor.*

But Eduardo needs an *emptor*, with or without the *caveat*.

Suddenly, the market is there.

Busy place, this market. A plaza of outside stalls set out under striped awnings, an old art deco building. The vendors behind the stalls look expectant or bored, the light around them striped as well.

He stops to survey the place, standing beside a pile of broken crates stained with grape juice, flies buzzing around. As he starts towards the stalls, a small flock of pigeons with red feet takes off in front of him, and then lands again almost immediately.

He threads his way through the crowds, through the noise of the market. *Ça, c'est frais?* the buyers say, over and over. Is it fresh? *Oui,* the vendors say patiently. *Oui, c'est frais. Tout frais. Très frais.*

He wonders whether one of them has ever lost his temper and said sarcastically: *No, they're rotten. See these bruises? I wouldn't buy them if I were you.*

But he has to admit their goods look fresh, unblemished, all these brightly coloured fruits and vegetables. Pints of blackberries, raspberries, gooseberries are arranged in mosaics, pale green corn cobs piled on a table, the tails of husk and silk carefully aligned. *Spéciale. Maïs du Québec. Très sucré.* Shallots hang in braids near persimmons, cucumbers are stacked up under a sign: *concombres contrôlés de façon biologique*.

An unofficial aisle of plants and flowers has sprung up, yellow lilies, basil pots, squash vines trailing and winding around the pavement, hibiscus, a tray of tired snapdragons, blue phlox on stakes. There are tables of artichokes, fennel, rapini. Where does all of this come from? He would be the first to say that he isn't much of a cook (true, says Geneviève) although he thinks of himself as at least acquainted with cooking, on speaking terms with it (really? says Geneviève). But some of this he has never come across before, some of it he has never seen.

Well, Geneviève is the one who usually cooks—he admits that. He hears her repeating the ingredients out loud as she looks for them in the cupboards, muttering to herself—white pepper, mustard seed, nutmeg—as if she is willing them to appear. It's simply a different type of science, she says—although this approach doesn't do much for the food. She is too fascinated by the chemical reactions involved, too carried away by the yeast flowering in warm water (a fungus, she says), too thrilled by the coagulation of the hollandaise—the end result, the taste is secondary.

'*Excusez*,' says a voice abruptly, loudly in his ear. He starts, looks around, and realizes that he has stopped in the middle of an aisle, blocking the crowds near a table of eggplants. He steps aside to let people by, and checks his watch. Concentrate.

He pulls out a small notepad and a pen, and makes notes while he walks, looking at the way people are moving past the stalls, where they talk to each other, where they avoid each other. He watches them hovering over sage leaves, fingering horseradish roots, putting garlic into paper bags. *Note*. He sees how they round corners, how they zigzag across the aisles. *Note.* If the market grew up gradually, did the way people move dictate the placement of the stalls? Or did the placement of the stalls shape the way people moved? Both? *Note.*

Not everyone is moving, though. He passes a man leaning up against a pole, a man with stringy hair, chapped lips, his jeans hanging off his hips.

'Du change, s'il vous plaît?' says the man, holding out an empty coffee cup.

Eduardo—prompted by what?—begins fishing in his pocket.

'Merci beaucoup,' says the man, while he is fishing. *'Il fait beau, n'est-ce pas?'*

He begins chattering away without waiting for an answer. Maybe the man is as much starved for conversation as anything else. Eduardo drops a few coins in the cup, and the man shakes it, and keeps on talking.

Eduardo shakes his head and moves on, and the man nods, resentfully polite.

This aisle is filled with conserves, pepper jellies, oils with sprigs of herbs in them, compotes. *Note.* A heavy woman in well-cut clothing, her face jowly, steps backwards and bumps into him. They apologize simultaneously. The heat under the awnings is stifling. *Note.*

A few minutes later, he sees the man with the stringy hair again—lighting a cigarette now, holding his hand around the match. He takes a pull on the cigarette, then turns to go into the market building. Not a bad idea. It should be cooler inside.

He follows the man into the building. And what a building—a gentlemanly old structure, now criss-crossed with unlovely ventilation ducts, external wiring, exit signs. Look at those bones, though, look at those arches, look at that high ceiling. Here we go, here we are. This is a building you could do something with.

Not everyone thinks of architecture as endlessly redeeming, says Patrick.

Idiots, says Eduardo sadly.

He walks slowly, examining the structure, the location of the pillars, the high windows. *Note.* The first floor has more fruit and vegetable stalls, the second, a *boulangerie* and butchers. He decides to begin with the second floor, and work his way down.

At the top of the stairs, he passes a counter piled with wheels of sourdough bread, stuffed *fougasse*, baguettes standing upright in a narrow basket. A few small tables and chairs have been set out for a makeshift café. This is where he is meeting the owner, but he still has a few minutes.

The crowds are lighter here, although still full enough to create queues. *Note.* At one of the butcher counters he watches a woman pointing to something in a glass case, raw and slick. The butcher whisks it onto a scale, and then wraps it in brown paper. The people waiting are studying skinned rabbits, containers of duck fat, trays of coloured sausages, grey, pink, green, dark yellow. He can almost see them weighing the choices in their minds, coming to decisions, on the brink of making purchases. The ham, not the pork roast. The chicken breasts, not the lamb chops.

There is a queue at the next stall as well where two countermen are working quickly, gathering up balls of cheese rolled in paprika, or twisted like skeins of yarn. While they work, they keep up a running patter with each other, with the people in line, handing packages over the countertop where bottles of vinegar, olive oil, pickled asparagus are lined up.

He can see a fish vendor towards the end of the second floor aisle, sweeping some trimmings from a cutting board with one hand. Young, stocky, a shaved head—he looks up and sees Eduardo watching him from a distance, with his pen and notebook. He turns back to his customers, but his movements become more self-conscious, more self-important.

The stall past the fish vendor is empty, the refrigerator counter unplugged. Trays of paint, brushes damp with solvent are sitting on a drop sheet around it. A painter is methodically rolling white onto the wall behind the counter. Even from where he is, Eduardo can smell the fumes.

He notices idly that the man with the stringy hair is in the throng ahead of him, almost abreast of the fish stall. The fish vendor looks up and sees him. He glances at Eduardo again, and then turns to the man.

'*T'a pas l'droit de fumer icitte,*' he yells at him. No smoking here.

The man with the stringy hair curses, and flicks some ash off his cigarette. He walks a few steps more to prove a point, and then tosses the butt of his cigarette away.

Less than a second.

Less than a tenth of a second.

A flash of light, so loud that its wake is soundless.

Eduardo finds himself on the floor, several yards back from where he was before. The scene in front of him has changed instantly. Fish are scattered everywhere, sole, whiting, snapper, their bodies draped over rubble, plastered to the side of a pillar. Prawns with tiny eyes are strewn along the aisle, as if they had suddenly felt an urgent desire to visit the cheeses. Several of the stalls are burning, and people are coughing, some of them on the ground like him, others still on their feet, clutching their bags, looking around wildly.

Then the shouting starts, the pushing, as people begin scrambling towards the stairs. The air begins filling with smoke and panic.

The crowd pushes around him, stepping over him. A man holds out his hand to pull him up, but Eduardo is too dizzy, nauseated, his legs too rubbery, and the man is swept away by the crowd.

The smoke is getting thicker, and he struggles to stand up again, holding on to a pillar. This time he makes it up, but is seized with a spasm of retching. Through the grey haze, he sees a woman bending over, the woman with the jowly face. She is pulling on something, swearing in French, trying to keep her large purse from sliding off her shoulder. He can see a work boot, a leg in an overall—she is pulling on the painter's legs, dragging him towards the door.

'Aidez-moi,' she is yelling, over and over. *'Donnez-moi un coup de main.'*

Eduardo tries to move in her direction, but he is still dizzy and has to grab a pillar to stop himself from falling. Other people move more quickly, a man who grabs the painter's other leg, a woman who picks up an arm, one of the butchers who takes the other arm. Together they form a clumsy, frantic group, lugging the painter—where? They disappear into the smoke.

Eduardo looks around, desperately trying to get his bearings—is that the exit over there? Over here? Over in that direction? He begins staggering towards the place where he saw the group disappear, fervently hoping that they knew what they were doing. A piece of flaming debris lands on him, and his jacket begins to burn. He tears it off, panicking when his hands are caught in the inverted sleeves, shaking the jacket frantically and then stepping on it to pull them out. He keeps on going, the smoke filling his lungs with lead, stinging his eyes. Then he almost trips over the fish vendor, who is on the floor, whimpering. He grabs for another pillar, and reaches over to pull the vendor to his feet.

Now he is moving forward again, coughing violently, but holding on to the vendor, half-dragging him along.

Over there—is that the red of the exit sign? Or the red of the flames? How can he tell? The flames are moving, the sign should be fixed. But this red is both. Horrified, he realizes that the fire is now between them and the exit. Is there another exit? He can't risk it—he has no idea where it is and he is almost suffocating now, his lungs straining for oxygen. He tries to move faster, but the vendor almost pulls him down. He yanks the man up again. Hold on, he tries to yell, but it comes out as a dry croak. Run, he yells into the man's face and then drags him through the line of flames, stamping on them, slapping at his clothes. Then they are blundering, half falling down the stairs.

Outside, his watering eyes make the light painfully bright. The plaza is filled with people, and the outside vendors are rushing to pack up before the fire spreads further. Over to one side, he sees the painter lying on a wooden loading pallet. He pushes the fish vendor over towards the pallet, and the vendor sits down heavily, dazed, holding his head.

A hundred cell telephones have sprouted, people with their heads tilted, their hands on their other ears. Some are calling other people, gabbling nervously in rushes. Others are calling the fire department, the ambulance service. *Oui, oui,* we know, we know. We're on our way, no need to keep on calling.

He clenches his fists to steady himself, but this is a mistake—the skin on his hands is beginning to smart with pain. The skin on his face as well, although his hands are worse. Other people are wandering around near him, looking with bewilderment at their arms where the skin is beginning to blister, feeling their faces and wincing. Some of them have lost their eyebrows and eyelashes, some of them have pieces of plastic bags melted on their wrists.

'Regardez,' they say, holding out their arms.

'Look,' they say, turning to other people, shocked.

An angular woman with a high forehead, frizzy hair is sitting on the curb next to the fish vendor, holding herself and rocking, smudges on her dress.

'Is my face burned? Is it bad?' she says in French.

The woman with the jowly face is standing near her, and turns around. She fumbles in her purse, and brings out a pocket mirror.

'Here,' she says, bending over, her jowls shifting. 'See for yourself.'

The angular woman looks into the pocket mirror, turning her face from side to side and touching it. Then she seems satisfied, and gives it back.

A man with burned eyebrows is holding out his hand for the mirror now. The woman with the jowls passes it to him, and he studies his eyebrows, rubbing his fingers across them. Some of the tiny charred hairs break off.

'S'il vous plaît,' says the man next to him, a man in a white apron, still holding a veal chop in one hand, as if he can't think of where to put it down. The jowly woman nods, and the mirror is passed over to him. He takes it greedily, examining the blisters puffing up on his skin, reluctant to the let the mirror go.

'There are other people,' the woman says sharply, pulling it away from him.

They are beginning to crowd around her, and Eduardo finds himself surrounded as well, although there is no pushing, no jostling—they wait patiently, silently for their turn. He steps away from them to make more room, to protect his hands from getting touched—the burning in his skin has become agonizing.

Fire trucks and ambulances begin arriving, their sirens winding down and then cutting off abruptly as they stop. Men jump out, opening doors, slamming doors, carrying equipment in a business-like way. The firefighters unroll hoses, trying to position a truck with a raised platform, while the ambulance attendants concentrate on rounding up the burned people.

Some of these people are moaning with pain, some are still dazed. The attendants speak to them matter-of-factly, intensely, gently, attempting to focus their fractured wits on moving towards the ambulances. The attendants touch them in unburned places, a back, a shoulder, to get their attention, to guide them. They hover over them, wrapping cold packs around their arms, their hands, and then giving them smaller packs to hold against their faces.

'Soon,' the attendants say, clearly, simply, in French and then in English. 'We will be at the hospital soon. See, wrap this around you. Here, hold your pack this way. We will be there soon.'

The firefighters are hosing down the building, the water leaping in wide arcs. Some of them have taken off their jackets, the reflective tape flashing in the sun. The crowds of people are hampering their efforts, getting in the way of the hoses.

'*Prenez votre temps*,' yells one firefighter sarcastically, as the police begin arriving. Take your time.

'*Ici, ici*,' calls an ambulance attendant, beckoning to Eduardo.

'I don't need help,' says Eduardo, although as he says it, he feels a wave of pain, as if someone had plunged his hands into boiling water.

'*Allons-y*,' says the attendant firmly, coming over to collect him. He gives Eduardo a cold pack, and begins leading him towards one of the ambulances.

Suddenly, there is a sound from the crowd—a drawn-out sound, part groan, part wonder, as if they can't help themselves. Eduardo stops and looks around.

The roof on the building is sagging in, flames surging through it. A brief hesitation—the crowd almost silent—and then the roof begins to collapse with a dull, cracking noise. A cloud of dark smoke, released, billows out into the sky.

All that food inside, burning. How strange, he thinks. The hanging slabs of meat, browning and then turning into charcoal, the baguettes turning into torches. And the fish, the clams and mussels yawning open before they become blackened lumps. The lemons shrivelling up, the tomatoes beside them bursting with small pops, spattering red pulp and seeds. Braids of garlic flaring up with a whiskery sound. Potatoes baking, red peppers wrinkling and then roasting, until they are nothing but ashes.

That *smell*, he thinks. What is it?

Something from his childhood?

Around him, he sees other people stopping, inhaling, even the ambulance attendants. Nodding their heads, inhaling again. Trying to remember what it is.

'*C'est quoi ça?*' someone murmurs.

Something vaguely familiar, but surprising, too, something stirring and tantalizing, like an itch creeping into his mind.

What *is* it?

'*Allons-y,*' says the attendant urgently, pulling him towards an ambulance.

The skin is an organ, says Geneviève. Epidermis, dermis, hypodermis. That's why it hurts so much.

Chapter Two

'Although I suppose it's really because skin is full of nerve endings,' she says a couple of days later.

Eduardo doesn't answer—not a good sign, she thinks. He can be abrupt or dismissive, but usually for a reason—not necessarily a good reason, but a reason. Since the fire, though, he seems strangely remote—look at him now, standing by the window with his gauze-wrapped hands at his side, watching the rain.

She puts down the viola bow, and comes over to stand beside him, hoping that maybe there really is something interesting going on in the street. No, he is watching needles of rain hitting the pavement below, dissolving at the second of impact. Hypnotic in its own way, but not something that would normally absorb him like this. Although it *is* wet—no doubt about that. Soaking, sopping, sloshing wet—water is flooding street gutters, spilling out of eavestroughs, and the street is full of umbrellas, brightly coloured against the dark grey of the sky. She watches a man with a newspaper over his head take refuge in a doorway, then take off his glasses and wipe them on his sleeve.

Give up, she says to the man silently. You're already wet. And about to get wetter. He ignores her advice—like everyone else.

Something about all this water, this splashing seems looser, freer, more—well, *fluid*, as if all these wet people might quietly sprout fins and start swimming down the street, their umbrella colours melting into them—blue and yellow stripes, iridescent greens, milky oranges, rose-pinks.

She shakes her head quickly. Eduardo is still mesmerized beside her.

'Fascinating as this is, I have to get going,' she says. 'Anything I can do before I go? Something to eat? Drink?' Something to think might be more useful.

He holds up his hands disgustedly—the bandages are boxing glove style. What can he eat? More soup in a mug, more awkwardly held sandwiches?

'I could stay home,' she says, a cheap offer, since she knows he will refuse it. Still an offer, though.

'Stop babying me.'

'It's not babying. You're injured. What's wrong with a little help?'

He says nothing.

'Well,' she says, 'just let me know when you want to rejoin the human race.'

He nods, as if this were a literal request.

She is half-exasperated, half-tempted to laugh. But there is clearly no point in staying around. Maybe he will be in a better mood tonight—day three since the fire. She puts the bow away, setting it into its velvet slot—the viola is already there—and closes the case. Then she stuffs some papers into a bag, looks around, and shrugs herself into a bright green raincoat. She pauses with her hand on the doorknob.

'You're not going to brood, are you?' she says, switching to English.

Stupid question, she thinks as the words are coming out of her mouth—of course he is going to brood. He is an expert brooder, a first-class moper—someone who settles down to it like putting on an old shirt.

'On second thought, maybe you *should* brood,' she says. 'In fact, now that I think of it, brood away. Mope to your heart's content. Although I suppose that isn't really possible—if you're moping, you can't be contented.'

'All right, all right,' he says. 'Go to work. I'm fine.'

But he isn't fine, he isn't fine at all, she thinks, as she takes the elevator down and emerges into the rain, joining the watery flow of people. A taxi? The rain has made them scarce, and she ends up waving futilely at ones that already have passengers. Where are her fins? Then a taxi pulls up to let someone off, and she runs over, splashing through a small lake at the curb, and slides into it.

The air inside is damp and chemically sweet—a cardboard tag with a picture of a lilac is hanging from the rear view mirror. This is competing with the smell of the driver's lunch, a half-eaten roti on the seat beside him. She realizes she is hungry, and has to restrain herself from asking for a bite. If only taxis served food—something small and quick. Maybe tapas? *Taxi à tapas*—an entrepreneurial opportunity. Not very likely, though, unless there are tiny kitchen facilities (and tiny chefs) that fit under a dashboard. Coffee? He could plug a tiny espresso maker into the lighter outlet. Also not very likely. The rain-snarled traffic is slow, and she settles back in the seat reluctantly, thinking about Eduardo.

Of course, he is morose from time to time. (You've really got to get that fixed, she said once. What? he said.

That tendency to gloom and doom. It's getting out of hand. He snorted.)

But the things he broods about (money, not enough work, money, his interns, money) are at least familiar—problems, yes, but not paralyzing, not upending. This, though—this is in a different category entirely. Obviously the explosion was shocking, naturally it would have been terrifying. She can understand him being stunned by it, unnerved by it for a few days. She is quite prepared to be particularly sympathetic, unusually kind, at least for a while. A long while, if necessary. But this new strangeness, this remoteness, is frightening her. Pull yourself together, she wants to say. Be a rattled Eduardo, be an irritable Eduardo, be any kind of Eduardo you want, but at least be Eduardo.

Is she too impatient? Give him time. But there is a tough little part of her, a second-youngest-of-six-children part, an on-her-own-early part, that finds this bewildering.

'Why can't he just snap out of it?' she says aloud.

The driver looks around, startled.

'Not you,' she says hastily.

He turns back to the wheel, his shoulders more guarded than before.

She looks out the window as the wet streets flow by.

You think you know this city? she said once to Eduardo. I guarantee you don't.

I've lived here most of my life, he said.

I've lived here all my life, she said, and I still don't know it, all of it.

An organism of its own. Complex, confounding, rich, bitter. A city of a thousand fissures, a thousand truces, of mercurial logic. What kind of city has its own glossary? *Francophone, anglophone, allophone*. What kind of city has its own mathematics? One that counts its inhabitants by languages, by mother tongue.

She wonders if human tongues have different shapes, depending on the language their owners speak. All those tawny-pink muscles—would they look different if their owners spoke a language where the r's were rolled? Or the l's were thick? Maybe she could create a taxonomy of tongues. A picture of a row of people, standing in a line, sticking out their tongues for counting floats into her mind—*francophone, anglophone, francophone*—and she almost laughs.

What about the ones who speak several languages? Would their tongues be particularly muscular? The body builders of the tongue world. She tries to remember what Eduardo's tongue looks like, but all she can think of is the feeling, his tongue on her tongue, his tongue sliding around her mouth.

When she speaks French to him, there is a second of hesitation on his part, a second to let her know that French takes more effort for him than English. Sometimes she gives in, begins lapsing into English to avoid this tiny resistance. Then she catches herself, and reverts to French.

A necessary evil, that vigilance. The rest of the continent is an English whale, capable of swallowing a language as easily as a sardine. *Les Anglais*, said her grandfather, her uncles, bitterly. The ones with our jobs, our money, our houses. Where did they find them? How did they get them? Why do they still have them?

No protests from her father, an anglophone, but so hapless he was almost inclined to agree with them. A man in a decades-long fight to the death with unemployment himself. Clearly someone was taking his jobs, too, whether it was *les Anglais* or not.

It's only a language, Eduardo says from time to time, impatient with the fervour, the controversy it provokes.

This leaves her almost speechless.

You English, she wants to say to him, with a twitch of anger. You're so smug. So dismissive.

But then again, he isn't really English.

And she is really—what? Someone surrounded by too much language—miles of nouns, oceans of verbs, mountains of conjunctions.

Luckily, there are fungi—non-talkers all of them, she thinks as gets out of the taxi at the university's biology building.

The walls of her third floor office are cluttered with photographs—indigo milk caps, *lactarius indigo*. Violet coral, *clavaria zollingeri*. Bitter oyster, *panellus stipticus*. Anemone stinkhorn, *aseroe rubra*. Witches' butter, *tremella mesenterica*. Swamp beacon, *mitrula paludosa*. Black jelly roll, *exidia glandulosa*. Scarlet cup, *sarcoscypha coccinea*. A gallery of fungal exotica, they make her feel at home.

But they're so unlikely looking, says Eduardo. So odd.

But so strangely beautiful, she thinks. So admirable. So generous.

Damp-users, earth-clingers, shade-lovers. Specimens that grow in marshes, on trees, in cracks. Slowly, silently eating their way through dead matter, releasing its elements into the environment. Reclaiming, transforming. Long may you gnaw. Tenacious entities, obstinate as—as what? As fungi. Look at those velvety textures, those glossy colours.

Who could possibly resist them? So appealing. So intriguing. Is it possible to love a whole taxonomic category? *You're the one that I want, darling Kingdom Mycota.*

The head of the department wanders in, a stocky man with reddish, sandy hair.

'Things going well?' he says in French. He is so shy that when he talks to her he looks at a point past her, down and to her left. At first she kept glancing in that direction herself, wondering what he was looking at, but now she is used to it.

'Yes,' she says, forcing herself not to react.

He is trying to hustle her along, pushing her to get her research done and published. What if someone scoops her? A serious problem—she worries about it, too, but his hovering only makes her tense.

'I'm really going as fast as I can,' she says, more calmly than she feels.

He looks vaguely dissatisfied. Despite his shyness, or maybe because of it, he is intent on taking the department to greater heights, goading his colleagues into becoming stars. He suspects Geneviève of irreverence.

You can't blame him for that, says Eduardo.

Yes, I can, she says. Besides, I don't like the idea that we should be staking claims to chopped up pieces of knowledge in career-convenient installments.

You might be in the wrong business.

The department head sniffs now. He often sniffs—he has allergies—but she is convinced that there is a kind of lexicon to it. This particular sniff sounds vaguely reproachful.

'Honestly,' she says, trying to sound sincere. 'I'll keep you posted.'

He wanders out again.

She waits a moment so that it won't seem pointed, and then closes the door. See, she says to him silently, I'm buckling down.

She sits down, rolling her chair closer to the desk. The floor has a tilt, just enough that she has to grab the desk sometimes to prevent her chair from drifting to one side, enough that she often turns around to find a file cabinet has slid open on its own. The department poltergeist—on the loose again. Now where would that be classified? In *Kingdom Monera*, with bacteria? No, must be *Kingdom Animalia*. But protozoa or metazoa?

From her window, she can see a scattering of buildings in the distance. Limestone—the building material of the French, according to Eduardo. And brick, the building material of the English. Even the physical landscape of this city is imprinted with language. (What do you mean, *even*? says Eduardo.)

She opens a computer file, puts up some charts. She is studying how particular fungi reproduce and spread on broken and disturbed ground. How their spores float lazily in puffs of wind, or shoot out from fruiting bodies. Most of them can reproduce asexually or sexually, through spores or fragments breaking off and starting new clumps. Sometimes they become shape-shifters, one form for sex, another for cloning.

How clever, how versatile. People aren't quite so lucky. Otherwise, children could be produced from hair, fingernails, even old skin cells. This would be an improvement over the present system, which seems so susceptible to malfunction.

It wouldn't be necessary, for example, to swim in a soup of drugs that stimulate human egg production, egg implantation. Eggs, eggs, eggs, she said to Eduardo, after one of these injections. Her breasts are swollen, her hair is wavier, her skin shinier, she feels slightly unravelled. Unravelled by what? Desire? Longing? Or just unravelled.

Is there a way to prevent all this from edging into other aspects of their lives, from leaving a thin film of discouragement on everything? If there is, she hasn't found it. The medical protocols are so pervasive that infertility hovers over them, invading their sleep, their meals, their conversations, the clothes they wear, the temperature of their bathwater.

Then, of course, there is sex, the piercingly delicious collaboration of skin, nerves, glands. This now revolves around arranging a meeting between his directionless sperm, and her artificially produced eggs.

We're just the dating service, she said to Eduardo a few weeks ago.

The pills, the injections, have become amulets for her, tokens of faith. She follows the protocols with the kind of dedication that is usually reserved for rituals, for sacrifices to petulant gods. Why? Because she needs something to follow. Because even petulant gods might be better than chance.

This doesn't make me a failure, does it? she says.

Impossible, says Eduardo obligingly.

Well, not entirely obligingly—if she is a failure, he would be one as well.

Is this one of his worries? Yes, in a way, but no, not really—she is the one with custody of this problem. He has enough of them as it is. She has a flash of him looking out the window, his hands in their bandages looking blind and numb. Give him time.

Although having him out of commission isn't going to help the baby project along. He may need time, but this is a time-sensitive endeavour.

I'm here to distract you, she had said a few days before the fire. He was sitting at their glass dining table, reading a report on a Sunday, and she slid onto his lap, facing him.

I have to get some work done, he had said.

Very impressive—in an unnerving, workaholic way, she said.

She bent her back and rubbed her face against his, a waxy, warm feeling.

What is this, anyway, that's so important? she said, sitting back and taking the report out of his hands.

It's an analysis of the fall of shadows from different angles of the mausoleum, he said.

Another of his projects. He reached for the report and she held it up over her head.

Just so that you know where the shadows are?

I'm using light and shadows to sculpt the building, to change its shape at different times of the day.

Very artistic, she said, as she tossed the report away, and started undoing the buttons on his shirt.

Very aesthetic, she said, twisting her body against his, a surge of heat rising in her groin. He sighed, and then grabbed her, rubbing his hands along the outside of her thighs, under her buttocks, pulling her closer to him as she moved her body.

In a minute, he had lifted her up into a sitting position on the table in front of him. He pulled up the thin fabric of her dress, and opened her thighs, running his hands along the inside of them. She heard herself make a small sound, something between a gasp and a moan, and he pulled the dress over her head, so that she was naked except for a small triangle of cloth that disappeared between her legs. He slipped his fingers under the triangle, and she reached for his belt, his zipper. Freeing his penis, she leaned over and took it into her mouth, tightening her lips, grazing it with her teeth. Then he stood up and pushed her onto her back, the glass table cold underneath her. Would it hold them? she thought, hazy and stupid with desire, imagining the table cracking and their bodies falling into a heap of broken glass. But then he was sliding inside her, and she didn't care whether the table held them or not.

Cracking? No, not cracking—pecking. She comes back to the present with a start. A bird is jabbing at something on the ledge outside her office window. Well, she should be working anyway.

She watches the bird bob as it attacks some tiny seeds. That bright rose head—a finch? *Class Aves* obviously. *Order Passerine* probably. Who knows the rest of it—family, genus and species?

Ah, the hopeless elegance of taxonomy—a valiant attempt to order the unorderable. Slotting every organism into its appointed place, whether it fits or not. Like fungi—misclassified as plants for decades, now with their own kingdom. But she is in favour of the effort, despite the futility. Go ahead, classify it all. Everything deserves a taxonomy. Everything? Everything. From organisms to—to orgasms, she thinks idly, still flushed with warmth. *Organisme, orgasme.*

Well, why not? In fact, maybe there is a taxonomy of orgasms already. But not something based on inherited traits, like biological taxonomies. Not something cold and clinical, either, measured in a lab. No, this would be qualitative, something authentic, from the point of view of the owner—so to speak. She could start a preliminary list. Out of the way, Linnaeus.

Paroxysmi arantiae. Orange orgasms. These start out gently, orange-scented around the edges, small and self-contained, a lacy little orgasm until a black tongue roars up the middle and the air cracks into shiny fragments.

Paroxysmi aestatis. Summer orgasms. These are warm, languid—nostalgic orgasms. They start out like arrowroot in a mild sun shower, then puff themselves up into milky, buzzing balloons, after which they slowly collapse, leaving a lingering sensation of uneasy calm.

A taxonomy would need real Latin, though—not one of her strong points.

She can hear the department chair's voice on the other side of her closed door (poor acoustic insulation, says Eduardo), talking to someone else.

Then she does buckle down, at least for several hours. Maps of spore patterns, wind currents, spreadsheets of data correlations. Not a particularly sensual activity, although not without its mycological charms.

After a few hours, she stretches in her chair. What was it she was supposed to remember? The thought is sitting in her brain, like a pebble, but only the idea of remembering something—not the something itself. It will come to her.

She leaves the biology building, looking up to see if it will rain again. The clouds are in fat grey swirls, as if someone has been ineptly finger-painting the sky, and the air is humid. More birds—finches again?—are strung out along the telephone wires. Strings, she thinks triumphantly. She needs strings.

The luthiers are in an old narrow building, instruments displayed under lights carefully aimed to bring out the patina of their wood, their curves and cavities. An antique Italian viola is set apart on a stand by itself, a cello nearby. Beside it are shelves of metronomes, polish, peg compound, rosin.

'*Cordes*?' says an elderly man sitting behind a counter, his neck scraggy, his knuckles enlarged. She is a regular customer, she plays on the weekends in an amateur quartet that performs in small places—gallery openings and salons. Although the first violin is always pushing them (nagging us, says the cellist) to play in bigger spots, find a broader audience, market themselves. (Delusional, says the cellist—we're nowhere near that good.) But they put up with him, the first violin, forgiving him on a weekly basis because his playing makes it worth it.

The elderly man stands up stiffly, and pulls out a tray with rows of strings, coiled in packages. He points out a new steelcore string, a premium brand.

'A nice, bright sound, particularly for a dark viola,' he says in French. He knows which strings she uses, but he always tries—almost perversely—to get her to try others.

She shakes her head. 'Too bright.'

Why else would she play the viola? The dark, rich sound.

He tries again—why does he do this? To show off his expertise? Or just to prolong the conversation with a fellow devotee, someone who understands the language? She talks with him for a minute, and then picks out the strings she always gets, two gut-core and two silver-wound synthetic. He shrugs, the skin underneath his chin wobbling. While he wraps the strings, she wanders around the store, looking at the instruments, the brazilwood bows, the carbon fibre cases. She doesn't need anything else, but she finds this place calming, with its craft, its dedication to the technical aspects of producing something so ephemeral.

In front of the Italian viola, she reaches out involuntarily to touch the side of the glowing wood. The cello beside it looks almost voluptuous in comparison. (Voluptuous? Too many hormone drugs in her system? But how much is too many?)

Although she knows people who are almost physically in love with their instruments. One of her friends, a cellist, mockingly describes it in terms of a love affair—first the initial rapture, the conviction that he is the luckiest cellist, that he has the finest instrument in the world. Then later a stage where his eye strays to other cellos, envious, until he finally gives into temptation, to a form of infidelity by actually trying them out. If only he had this instrument, he thinks, his playing, even his life would be so much better. And finally, the reconciliation, the bittersweet gratitude for the one he has. Until he acquires a new instrument, and starts all over again.

'And how will you be paying for this?' says the elderly man.

Outside, she runs into Patrick, coming out of the Métro.

'How are you?' she says in French, laughing as they untangle themselves. She is surprised by how happy she is to see him.

'Good, good,' he says.

'No, really, how *are* you?' she says. She regrets this as soon as she says it, the sappiness of it. She was thinking of his divorce, but the words are suddenly wrong.

'Well, not that good. Not that good, really, but we've signed the papers so I think the worst is over. Although I keep saying that.'

He has an agile mouth, a narrow face—the face of someone capable of surprising generosity, small bitternesses. Even in repose, his expression often suggests he is listening intently to something. An occupational hazard, he says, although this can't be true—otherwise all lawyers would look like that. (Notaries, says Patrick.) Notaries, then, but the Québec version—a lawyer without a courtroom. He is more of a talker than a listener, anyway, at least in terms of quantity—sentences roll out of him in effortless streams, swirling around him. If she feels swamped by language, she can blame him in part, generating it out there by the yard.

'What's new on the fungi front?' he says. 'Any fungi breakthroughs? Any insights into the secret lives of plants?'

'Not much,' she says. 'They're not really given to dramatic developments.'

'Although they're not really plants, either,' she adds. 'Genetically, they're closer to animals.'

'You're kidding,' he says, impressed. 'And there hasn't been a horror movie about this? The Night of the Living Fungi? The Fungi of the Opera? Revenge of the Fungi?'

'Not that I know of. But they not very mobile. In other news, though, Eduardo was caught in a fire.'

'What happened? Was he hurt? Is he all right?'

'More or less,' she says. 'He has some mild burns on his face, serious second degree on the backs his hands.'

'I bet he's taking that well,' says Patrick wryly.

You bet right, she thinks. Although even you, his closest friend, might be surprised at how—how what? How distant he is. How brittle he seems.

'Come for dinner tonight,' she says, on an impulse.

He looks at her curiously.

'Sure,' he says. 'Sounds good. What time?'

She takes the stairs quickly down to the platform of the Pie-IX station as the train rushes in. Inside the train, she finds herself a seat and leans her head against the window, closing her eyes. Next stop, Joliette, a familiar blur of pale yellow tiles. Joliette, Cuvillier, Aylwin, streets of a disorganized childhood—a blur of laminated furniture, crucifixes, old cooking smells, a discarded stove sitting at the curb. Steel wool pads in the sink, turquoise soap foaming out of them, the headache-sweet taste of port, tiny cups on New Year's Day. *Bonne et heureuse année.* A string of landlords, a widow who had inherited several buildings, a sullen Hungarian who had given himself a French name. Ashtrays everywhere—glass ashtrays, tin ashtrays, jar lid ashtrays—smoking was so pervasive, so natural that each child took it up in turn as a teenager, welcomed into the loyal order of nicotine. The family landscape was criss-crossed with alliances on the basis of cigarette brand, on the basis of borrowing and bargaining—a web as strong as anything else.

She was an apostate though, a non-smoker, escaping outside as much as possible. Even in the winter, she was outdoors so much that the lining in her snow boots would wear thin and tear, then crease into ridges under her feet until she pulled out pieces of it. Her mittens were full of ice-clumps, crystals hanging off the wet wool, her legs always cold and red under her leotards—she would never wear snow pants, snow pants were evil, designed to make the wearer look babyish.

But when the snow melted, they chanted songs as they hopped around the sweep of a jump rope. *Allô, allô, allô, monsieur. Sortez-vous ce soir, monsieur?* Are you going out, sir? *À la salade, je suis malade. Au céleri, je suis guéri.* With some salad, I am ill. With some celery, I am cured.

Cured of what? said her father once, shaking his head. An unlucky man, blunt-bodied, bright-eyed, sardonic. One of the headstrong—one of the lunatic—school of lovers who pop up from time to time, determined to marry someone across family lines, language lines, religious lines, land mines. Alas, that was his peak, his shining moment of courage (or romance? his shining moment, anyway)—no match for the steady drip of chronic unemployment. When he was winning the battle of the jobs, he was a shipper-receiver, a forklift operator, a bartender, a rodman on a survey crew, an inventory manager, a sprinkler system installer, a maintenance man, a short order cook, and a driving school instructor. Which only means he's bad at everything, says her sister, the eldest, Marie-Thérèse. His only talent—a knack for failure.

But is that fair? thinks Geneviève. No. Look at the efforts he made, look how hard he tried. If ever there was a man who should have received an A for effort (or at least a B), it was him. True, he turned his efforts into something of an art form—they seemed to loom larger than actually finding employment, as if they were the real focus. But maybe this was because they yielded so little success to balance them out.

And he made efforts with other things as well—winning over his wife's family, for example, where he earned an honorary membership by recasting himself as a kind of Francophone convert. He had learned a substantial amount of French in his various excursions in the labour market, although his vocabulary was skewed as a result (if you ever need someone to tell you how to install a sprinkler system in French, he's your man, says Luc, her youngest brother). But he was accepted in a way that a luckier man might not have been, a fixture at family parties where they ate pork stew, sponge cake with thick white icing. He and her uncles, garrulous, flushed with beer—sometimes they would take the top off an ironing board, and set it on the ground. *Couche-toi, couche-toi,* they said—how old was she the first time? Four? She lay down on the board, then they blindfolded her, and tied her on. Then her father made high-pitched airplane noises, while they shook the board, while the board took off in their hands, swooping and dipping through the air. At the last minute, the rope was undone. *Au secours, mayday, mayday*, they yelled. She rolled off the board, falling, shrieking, to find out that she was only a few inches above the ground.

Encore, encore. No, it was another child's turn.

At the end of the evening, her father and uncles would go out to warm up the cars, to shovel the snow off the outdoor staircases. Then they carried the sleepy children down, and settled them into back seats. She can almost feel the cold air outside, then the musty warmth of the car heaters roaring in front, the comfortable feeling of being squashed in with other small bodies in snow suits.

The Métro train jerks, and she looks around—they are pulling into Papineau. A mural flashes by, brightly coloured figures. Too stylized, simplistic for her.

But *les Patriotes*, says Luc. You don't have an eye for art. Or a feel for history.

A doomed rebellion? she says.

It's educational, he says.

Yes.

The train starts up again and rounds a corner sharply. Then it begins slowing down, even though they are between stations, until it has stopped completely. The windows are black, and the lights in the train car give it an artificial intimacy. The other passengers begin looking up from their newspapers, their phones, frowning, or looking anxious. They eye each other surreptitiously, wondering whether they will be trapped, assessing the people around them—are they useful? Are they dangerous?

The lights flicker out, and come on again. Then the train starts up, pauses and begins picking up speed.

The unfathomable mysteries of the Métro. She should be thinking about making dinner instead, now that Patrick is coming over. She tries to remember what they have in the house, to think of something interesting to cook.

Something that transforms from one state to another. Or that involves engineering. Something stuffed? Crusted? Whipped? Is there anything in particular that Patrick likes? She can't remember.

As she walks into the apartment, the telephone is ringing.

'*Allô-toi*,' says a voice. Luc himself—conjured up. Be careful what you think about. Or who you think about.

The brother closest to her in age, a partner in childhood crimes, rowdy, clever. The one with unlikely schemes that could work, that almost work, but never do. Right now he manages a bar on rue Ste. Catherine, a bar with unknown art for sale on the walls. And a shelf life of a year and a half, said Marie-Thérèse.

But this is who he is calling about.

'She fell off a table, changing a light bulb,' he says in French. 'She's broken her left leg in two places and several bones in her right foot. They've put pins in the broken leg, but she can't walk with both legs out of action.'

'Poor her,' she says, meaning it. Marie-Thérèse, heavy-hipped, energetic, eyes outlined in black. Maybe she tries to order them around too much (no *maybe* about it, says Luc), but she was stuck with this job by birth order. And she had to help deal with her father, a man who was tenderly treacherous when he had too much beer in his veins. She is a good person, if *good* means warm, brassy, loyal, a respecter of conventions. And maybe this is what it does mean.

'Poor us. We have to divide up the kids,' he says.

What? Not that good.

'At least for six weeks, until the cast comes off. Jocelyne's taking the baby, Raymond and his wife are taking Émilie. Can you take Matty?'

Marie-Thérèse's husband disappeared a few years ago, his only presence occasional calls from greyhound racing tracks in Florida, promising money that never arrives.

'I have to work,' Geneviève protests. 'And we're going to Portugal in a few weeks. And I have a concert coming up. What about *maman*?'

'She has the flu. And she can't keep up with Matty, anyway.'

'Well, what about you?' she says. Let's be egalitarian here, an equal opportunity to be useful.

'We have the girls, and I work, too.'

Matty, she thinks, a four-year-old on the loose, a squirmer, fidgety as a drop of mercury. Gregarious, too, a born visitor, someone who makes the rounds in Marie-Thérèse's apartment building, knocking on doors, inviting himself in. Trading on his four-year-old charm, until his mother retrieves him, embarrassed.

'I'll have to talk to Eduardo,' she says, looking around. He must have gone to work—no surprise there. Although what could he possibly do without his hands?

'We're desperate,' says Luc. 'We can take him when you go Portugal. We can take him for the concert. But he's a handful, he's driving us crazy, as if the girls weren't enough.'

What about Paulette? But her other sister lives in Hawkesbury, and has four children herself.

'I thought you'd want him,' says Luc. 'You, of all people.'

Don't say it, she thinks. Stop right there. Don't even consider putting your clumsy words down on that raw spot.

'Are you still there?' he says.

'All right,' she says quickly. 'But give me another day. I'll pick him up on Wednesday.'

After she hangs up, she sits there for a minute.

What else could I do? she says to Eduardo, already rehearsing.

And who knows, maybe this will help, maybe this will be a distraction. Maybe Eduardo will snap out of it.

Or maybe he will just snap.

Half an hour later, she is cutting up cold beets and peppers for a salad, the beets staining the cutting board, the counter, her hands. It looks as if she has been murdering someone. Will the beets and peppers be too pungent together? Probably, especially with vinegar and hazelnuts. Maybe Eduardo and Patrick will be too distracted by the colours to notice their tongues drying out. The hazelnuts are stale, anyway. They always seem stale to her, although perhaps she is too selective, too picky about this. Every year, her mother brought home hazelnuts from Rimouski, where her aunt and uncle had a farm. The green nuts were bitter—she and Jocelyne tried them once, and spat them out. But her mother would put them in a cotton bag, and bang them against the wall to crack them. Then she would let them dry out, until they changed from green to gold, until they tasted buttery.

She looks at the time—Eduardo should be back soon, and Patrick will be here shortly. The rest of the dinner is a goat cheese soufflé, the best she can do on short notice, especially now that half of her brain is thinking about Matty. The half that isn't thinking about Eduardo. At least the soufflé will puff up in a satisfyingly metamorphic way. And Eduardo can eat it with a fork—no knife required.

Stop babying me. But what would *adult*ing consist of? It sounds vaguely erotic—maybe she should be wrapping herself around him, like some kind of naked bandage.

They see Patrick often, of course, but before it was usually with Beth. Small, mouse-boned Beth—her straight blonde hair tucked behind her ears, blue-white skin, her legs undefined, almost calf-less. Not merely seven-eighths of her below the surface, but nine-tenths. The divorce has meant she has been swamped in emotion, something her carefully assembled self is unable to handle. She seems to hate Patrick for this exposure of herself as much as she hates him for leaving her. She has turned into one of the Furies, but a forlorn Fury, small and inconsolable.

How many kinds of hatred are there? She's working her way through the whole set, says Patrick despairingly.

Classes of anger, Geneviève thinks. *Kingdom Sensus. Genus Odium, Species Furore.*

Most of Beth's hatred is reserved for Patrick, although Geneviève and Eduardo have attracted some by association. She refuses to speak to them, to respond to their tentative overtures. Geneviève is unsure of what to do, but she is temporarily stymied, even a little hurt herself by Beth's rejection.

We've been fired as friends, she says to Eduardo.

Patrick can be angry, too, but he has an opaque quality about him—someone who has several additional layers between himself and the world. He often seems as if he were on the verge of being cautious and amused at the same time. Dry, hidden, despite his volubility. In a word—English. Not the real English, of course. The real English includes her father, who, far from being hidden, always seems to be in the middle of an audition for an audience that isn't interested.

She slides the soufflé into the oven, and opens up a bottle of wine, a new one. It has a harsh aftertaste, but she sits down with a glass anyway, and tries to think of a way to explain to Eduardo that she has landed them with a four-year-old whirlwind without consulting him.

His relationship with her family is complicated enough. He still has the status of a resident curiosity—they haven't entirely digested the fact that she has inexplicably linked herself to one of *les autres*. Having a father who was English was messy enough as it was—but a Portuguese now, too? Although they are too sophisticated to put it this way, or even to feel it this way. With them it takes the form of being genuinely perplexed—in varying degrees—as to what she sees in Eduardo, the person. Not that they have been unwelcoming, quite the opposite. Marie-Thérèse in particular is self-consciously friendly. She is a little over-stated about this, though, as if Eduardo might be tone-deaf.

Sometimes she talks to me as if I'm a beginner she's encouraging, he said once. As if she's bringing me along, and has been pleasantly surprised by my progress.

I think it's an oldest child reflex—she's like that with us, too, Geneviève said. Although this isn't entirely true.

Let's keep in mind your family is not exactly enthusiastic about me, either, she added. How do you say *les autres* in Portuguese?

Although maybe that should be: how do you say *infertile* in Portuguese?

Then Patrick is at the door.

'*Salut*,' she says, as she kisses his cheek. He is more subdued than usual—he seems to be listening carefully to her kiss. She is suddenly aware of his hand on her arm, his smell—what is it? Something fresh and bitter, like parsley? Does he cook?

'Dinner is almost ready, all we need is Eduardo,' she says in French, more heartily than usual. Too heartily, she thinks, wincing internally.

She pours him a glass of wine, and they sit down. Almost immediately, he is talking about the divorce again, as if he is continuing to answer her question from the afternoon. But there is no doubt that he is surprised by the divorce—not the idea, since it was his idea, but how overwrought and crazed it has been. His only experiences before were the couplings and uncouplings of his twenties, when people came and went with relative ease. The craziness is Beth's fault. Of course he would say that, Beth probably has a different story. But he blames her so subtly, so charmingly, so kindly that it is almost impossible to think otherwise. (Although maybe this is why Beth is so crazed? she thinks, but only for a second.) Then there is Chloe, their two-year-old, who has become withdrawn, bewildered.

'What a thing to happen to a child—she seems so confused. I don't know what to do with her.'

He pauses for a minute.

'I'm thinking of taking my nephew to the insectarium on Saturday,' says Geneviève. 'Do you and Chloe want to come?'

She knows this isn't what he means, but says it anyway.

'These things are so difficult,' she adds weakly.

'And the worst thing is that you end up boring your friends. Let's talk about something else,' he says.

He gestures towards the viola case.

'How about some real entertainment? I don't think I've ever heard you play alone, only with the group.'

'I'm better with the others,' she says hastily. She is self-conscious about performing for an audience of one, at least an audience of one that isn't Eduardo.

'I don't believe it. Anyway, if you don't play, I'll start talking divorce again and then you'll be sorry.'

'I'll have to change one of the strings.'

'Go ahead.'

She finds the new package of strings, and takes one out. Then she threads it through the tailpiece, over the bridge and into the peg hole. She tightens it up, and then tunes it quickly.

'Well, this is a viola,' she says finally, smiling at the absurdity of this. 'As you might say, Exhibit A.'

'Exhibit A,' he says solemnly.

'And this, this is the bow. Exhibit B.'

'Exhibit B,' he says.

'You draw Exhibit B across Exhibit A.'

She draws the bow, and a warm, liquid note comes from the instrument. The shoulder of her bow arm twinges. Bursitis—the musician's plague. She plays a fall of notes, dark sounds, sounds that make her think of oak bark, of damp earth.

'Then the sound waves go to the bones of the inner ear,' she says, holding her ear lobe. 'Exhibit C.'

'Exhibit C,' he says, holding his own earlobe in imitation.

'The hammer, the anvil, the stirrup,' she says, laughing.

'They sound like constellations,' he says. 'Like the big dipper, or the archer.'

He is a small-time stargazer, the possessor of a telescope, some charts, and not much else.

'They're too busy to be constellations,' she says sternly. 'They have things to do. They can't afford to hang around in the sky looking decorative.'

'Stars aren't decorative,' he protests. 'Well, I suppose they are in a way, but that's not their main purpose.'

'They have a purpose?'

'I was speaking generally,' he says with dignity.

'Your turn,' she says. 'Speaking generally.' She hands him the instrument, and gets up so that he can sit in the chair.

'Hold the bow like this,' she says, showing him, and then handing it over.

He draws the bow several times, but the notes sound sour.

'Move the bow lower, and make the stroke bolder.'

The notes are a little better, but not much.

She stands behind the chair, and puts her hand on the bow, around his hand to show him, producing a few swift strokes. His hand is unexpectedly warm and dry—she is surprised by how good it feels.

'Now try it again by yourself,' she says quickly.

'No, no, this is a little excruciating. I think I'd rather hear someone who can actually play,' he says, getting up and handing back the instrument.

She sits down again, and plays a series of notes. Then, without thinking about it, she moves into a piece she is working on. This time the notes are throaty, amber, although she hears the flaws, the imperfections. She plays for several minutes, absorbed—oblivious—and then stops.

They are silent. People are talking in the hall outside the apartment. Their voices swell, and then fade away. She can hear the muffled sound of a chair scraping in the apartment above, water draining down a pipe in the wall.

Keys clink at the door, and they start.

'Patrick,' says Eduardo, surprised. He is balancing an armful of papers between his hip and his arm, avoiding his hand. He sets them down awkwardly on a side table. 'Did I know you were coming?'

'No,' says Geneviève. 'It was an impulse.'

'Well, good,' he says in a distracted way. 'Good impulse.'

'Pour yourself some shoe polish,' says Geneviève, putting her bow down, 'while I get things ready.'

She settles the viola back in its case, and disappears into the kitchen.

'Is it really that bad?' says Eduardo, looking at the bottle.

'Yes,' says Patrick. 'But fortunately, it has alcohol in it. Here, let me do that.'

He pours Eduardo some wine, and Eduardo cups his hands clumsily around the bowl of the wine glass.

Fifteen minutes later the soufflé is done—good timing, she thinks, congratulating herself. She brings it out with the blood-red salad and they settle down to eat. The soufflé is handsome-looking, but too runny inside. She looks at it dubiously.

'The recipe said it would be a little runny, but not like this.'

'All good soufflés are like that. In fact, this is probably the epitome of souffléness,' says Patrick gamely.

'You're a bad liar, but an excellent guest,' says Geneviève.

They eat it anyway, working around the runnier parts. Eduardo fumbles with a fork for a while, and then gives up.

'So tell me about this fire,' says Patrick easily. 'How did it start?'

'Not much to tell,' says Eduardo. 'A man in front of me threw a lighted cigarette away, and it landed around some painting equipment. I don't even know if it landed in the solvent, or just ignited the fumes.'

'*Res ipsa loquitur,*' says Patrick, pouring himself some more wine.

'Very erudite,' says Eduardo dryly. 'What does it mean?'

'The thing speaks for itself. It's a legal doctrine about causation—the legal version of cause and effect. You'd like it—it's full of bits of philosophy.'

'If the thing speaks for itself, why do you need a doctrine to say it?'

'I refuse to answer that,' says Patrick, 'even though there is a perfectly good answer.'

'I wonder whether the fish vendor would have yelled at him, if I hadn't been there with the pen and notebook,' Eduardo says, almost to himself.

'Now you're into another doctrine,' says Patrick. 'That's the *but for* test, another handy little attempt to categorize cause and effect, to sort out proximate causes.'

'As in *but for* Eduardo, this wouldn't have happened?' says Geneviève.

'That's the idea,' says Patrick. 'Except that *but for* the man throwing away a burning cigarette, it wouldn't have happened. Or *but for* the painting going on during market hours, it wouldn't have happened, either. '

Eduardo shoves his plate away.

'Where did the perpetrator end up, anyway?' says Patrick.

'No idea,' says Eduardo shortly. 'I didn't see him after the explosion.'

'I suppose if he wasn't hurt badly, he might have taken off. Afraid he was going to be blamed,' says Patrick. 'But look on the bright side. It's convenient if you're going to be renovating it.'

Yes, let's look on the bright side, thinks Geneviève.

'Hard to tell. It depends on what's left, how weakened the structure is, what the renovations will be.'

'No, I mean the owner should have some insurance money now,' says Patrick.

'I guess that's true,' says Eduardo reluctantly. 'But he isn't returning my calls.'

'Let's face it, you don't like the idea of buildings burning down on general principle,' says Geneviève, getting up to clear their plates.

'No,' says Eduardo heavily, 'I don't.'

'Coffee?' she says. 'The tart will be a minute.'

Patrick follows her into the kitchen with some cutlery. He jerks his head towards the dining room, and raises his eyebrows.

She grimaces in response, while she takes out a peach tart and puts it into the oven to warm it up.

As he walks out again, she hears him say 'So what else are you working on these days?' and she feels a spurt of gratitude.

When she brings in the coffee and the peach tart a few minutes later, they are standing over some drawings spread out on the table, and she cranes her neck to see.

The drawings are ink lines with some pale splotches of colour. They show a large, three-story building from several perspectives, with stone rectangles, a sweep of glass. Full of cold, brilliant lines.

She is struck by the flow of it, a kind of physical cadence.

If architecture is frozen music—who said that? Goethe?—is it possible for it to be unfrozen? Turned back into music? Interesting thought.

Beside her Patrick exhales, a small sound, as if he had just bitten into something indescribable.

'How do you do this?' he says.

Eduardo hesitates.

'You mean the process, how it develops?' he says.

'I'm not sure what I mean,' admits Patrick. 'What is it, anyway? The building.'

'A mausoleum,' says Eduardo, rolling up the drawings again, tapping them into a carrying tube.

'All that, simply for dead bodies?' says Patrick.

Perfect, thinks Geneviève. Another cheerful topic.

'It's very preliminary,' says Eduardo. 'I don't even know if they'll like it.'

'The bodies? I doubt they'll have an opinion,' says Patrick. 'Everyone else should like it though—it's stunning.'

Geneviève passes around the coffee cups, and begins cutting the tart.

'Did you bake this, too?' says Patrick.

'No, you're safe,' she says, handing him a plate. 'I picked it up on the way home. It should be entirely edible.'

She ignores his polite protest, and runs her finger along the side of the cutting knife to pick up crumbs and peach, then licks her finger.

'Luc called today,' she says casually. She knows she should wait until they are alone, but she doesn't want Patrick to leave. Why not? Perhaps he is a buffer? Yes, but more than that. He is a breath of normalcy, he seems to lighten and balance things. Besides, Eduardo doesn't care that much about social graces—he is perfectly capable of erupting in irritation in the presence of other people.

'Marie-Thérèse broke her leg and a foot, and they need us to take Matty for a while.'

'All right,' says Eduardo absently.

All right? All right? Was he listening? This is worse than she thought. This is so completely, so profoundly unEduardo-like. *You of all people.*

'I'm going to watch the news,' he says abruptly. 'See if there's anything else about the fire.'

They take their coffee and plates into the living room, where she flicks on the television for him—the buttons on the remote are too close together for his bandaged hands.

It seems odd—almost intimate—for the three of them

to be sitting around like this, not talking, instantly stupefied by the television. She and Patrick start speaking at the same time, and then stop—a reporter with a microphone and a staccato voice is talking in front of the blackened market building. Several official-looking men are poking around in the debris.

'*Suspicious*?' says Eduardo, his voice rising. 'They're treating it as suspicious?'

'Suspicious?' says Geneviève blankly.

'That's absurd,' says Eduardo. 'I saw the man throw the cigarette myself.'

Chapter Three

'Now it's arson?' says Patrick in French. 'You seem to be having an eventful week.'

He swings two-year-old Chloe up on his shoulders, where she cups her hands around his face.

'That's what they're saying,' says Geneviève. 'Or suspected arson. I guess that's an upgrade from *suspicious*.'

They are standing in the middle of a greenhouse, on their way to the insectarium—the three of them and Matty, who is spinning around, trying to make himself dizzy. Geneviève has negotiated this detour to see some fungi.

Just a quick look, she said.

Insects, fungi—it's all the same to me, said Patrick. All part of the unpleasant conspiracy of nature.

In fact, he is feeling unexpectedly self-conscious. Surely he and Geneviève have spent a great deal of time together, what with one thing and another? Maybe they haven't gone anywhere alone before—he can't remember. But then, he doesn't usually have Chloe for the weekend by himself, either—it's a brand new post-divorce world. And here he is in a greenhouse, of all things—not something he has any interest in. This particular area isn't even much like a greenhouse—instead the light is low, the air damp and warm. Be a good sport, he says to himself, and bends down to read the signs stuck in the plant beds.

Magpie ink cap, *coprinopsis picacea.* Orange peel, *aleuria aurantia.* Black trumpet, *craterellus cornucopioides.*

'They're just so *ancient*,' says Geneviève, sighing happily.

'Huh,' says Patrick, looking at the odd shapes, momentarily speechless. Obviously an acquired taste.

'Can you imagine? They've existed for about 400 million years,' she says.

'Old,' says Patrick. Then he yawns.

'Sorry,' he says, 'I've been having trouble sleeping.'

She certainly loves this stuff, he thinks, looking at her glowing face. He wishes he felt that way about his work. Or anything else.

He wonders what it would be like to be with someone so animated, so alive. Beth was cool and smooth as milk glass, almost mesmerizingly poised—that is, until she wasn't, until she shattered. Geneviève is warmer, more impulsive—at the moment, she seems to be emitting a kind of fizziness, as if she were so delighted to be around the fungi that she couldn't keep it dammed up. Before, he would have found this exhausting. Now he wonders whether it might be possible to acquire this kind of enjoyment of things—in things—from someone else? Would it rub off on a friend? A husband? A lover? It doesn't seem to have rubbed off on Eduardo—he could hardly be less effervescent. Particularly now.

'Pig's ear,' he says, reading a sign in the earth bed. *Gomphus clavatus.*

'Ear, ear, ear,' Matty chants as he whirls around the path between the beds.

'Look at this one,' she says, pointing to a model. 'Devil's cigar. It only grows in Texas and Japan. It's the Texas state fungus.'

'Really? A state fungus?' says Patrick incredulously. 'Well, I suppose if any state had one, it would be Texas.'

He bends over to look at it, holding on to Chloe's ankles. The air is damp and earthy-smelling.

Matty stops whirling, staggers and sits down on the path, collapsing into a cross-legged position, a grin on his face. Geneviève pulls him up, and they wander over to look at a map—a wooden chart of the different greenhouse areas, and the various outdoor gardens.

'How about the Pavilion of Infinite Pleasantness?' says Patrick, studying the map. 'It's in the Chinese Garden. I could use some infinite pleasantness.'

'Or there's the Tower of Condensing Clouds.'

'I think I've had enough clouds, condensing or otherwise. But there's a Garden of Weedlessness, too. Now that sounds like my kind of garden. Unless there's a Garden of Gardenlessness, in which case, that would be even better.'

Chloe begins kicking her heels against his chest. He catches her feet for a few seconds, but when she protests, he swings her down off his shoulders.

She runs over to where Matty is sticking his fingers into the loam near a clump of puffballs, the size of white balloons. Geneviève moves quickly to rescue the puffballs, but Matty spins off again, leaving a small trail of earth. He bumps into a man with a long nose, and ends up on the ground again.

'*Excusez,*' says Geneviève, pulling Matty to his feet again, and dusting him off.

'*Excusez, excusez,*' Matty chants as he is being dusted.

'That kid is going to get into real trouble someday,' says Patrick. 'I think we might have to skip the Pavilion of Infinite Pleasantness.'

He lines them up, Chloe, Matty and Geneviève.

'We have an important mission today. Nod your heads if you understand,' he says sternly.

They all nod their heads, although Chloe and Matty have no idea what he is talking about.

'Our mission is to go to the insectarium and find out the answer to a question. Nod your heads if you understand.'

They nod their heads again, although Chloe is now nodding vigorously from side to side, rather than up and down, and Geneviève is laughing.

'The question is this: which one is the biggest, hairiest, ugliest spider in the place? Do you accept this mission?'

They nod solemnly.

'Follow me,' he says. 'Or actually, follow Geneviève.'

She leads them outside, down the path to the insectarium, and he brings up the rear. From this position, he can't help noticing how she moves—as if her joints were coated in something soft. Has she always been like that? For some reason, she seems more—more what? More noticeable than before. More tangible. More something.

Watch it, he says to himself.

No harm in watching, he says to himself.

'Here we are,' says Geneviève.

The insectarium is large and airy, a building with white wings.

'Impressive,' he says, looking around. 'I had no idea insects live so well.'

'You've never been here before?' she says. 'It was actually established by a notary.'

'One of the brethren,' he says. 'Well, there you go. It must have been the focus on minutiae. Insects are the obvious fit.'

They wander through a room with horizontal glass cases on stands, chartreuse walls, and deeper green columns flared out at the top. Another room has wall cases with beetles in jewel-like shells, masked butterflies in rows. Matty is enraptured, his hands flat on the glass, his nose an inch or two away. Chloe is on Patrick's shoulders again, peering over beside his head.

'Remember this,' says Patrick to her. 'Nature is truly bizarre.'

A group of older children begins to fill the room, the children chattering in English, poking each other, jostling each other.

'Look at that,' they say. 'Look at this. Look over here. Stop pushing.'

'Look at *this*,' says Patrick. 'Monarch caterpillars eat their skins after they shed them. Seems a little on the cannibalistic side.'

'But tidy,' says Geneviève. 'You have to give them that.'

Matty is now fascinated by the other children instead of the insects.

'Salut,' he says excitedly, jumping up and down. He doesn't speak English.

The older children ignore him, they are busy talking, pushing each other. Only one boy notices him, one of the taller ones. He gives an almost imperceptible wave, barely raising his hand from his side.

'Good to see there's some civility left in the world,' says Patrick.

'He probably has a younger brother,' says Geneviève.

'The specific overrides the general.'

She turns to him.

'It's a rule of statutory interpretation,' he says. 'A specific provision in a law takes precedence over a general one. In this case—an individual over the general group dynamics.'

'Of course,' she says, laughing. 'You took the words right out of my mouth.'

'Are you mocking me?' he says sternly.

'What does it sound like?'

'And to think how nice I was about the fungi.'

'Fungi are entirely—absolutely different,' she says loftily. 'Fungi *deserve* respect.'

'No-one could accuse you of a lack of focus.'

'You'd be surprised,' she says.

Look how well we're doing, thinks Patrick.

Look how hard we're trying, thinks Geneviève.

'Arson,' says Patrick out loud, sitting at his desk several days later. This is not an area of law he knows about, but it doesn't seem like the right charge—wouldn't some kind of intent be required? Criminal negligence maybe, but arson? Of course, the man might have known about the painting and the fumes, might have planned to throw the cigarette butt anyway, with the fish vendor's reprimand simply a handy excuse. Maybe this was a particularly clever form of arson—no need for a nighttime raid with a can of gasoline.

Or maybe the man was just careless. This seems more likely. Of course, they haven't found him yet, but human ineptitude seems to be more common than malfeasance—at least if his clients are any example.

He glances at the rest of his appointments for the day—girding himself for more of the usual parade of misery.

Although this is a little hard on his clients. Many of them are reasonable people, stumbling through circumstances beyond their control. But sometimes the sheer weight of them, all their problems added together, seems overwhelming.

Clients, from the Anglo-French *clyent*, from the Latin *cliens*. They collect outside his office, they call, they leave messages at all hours. The messages are pleading, or angry, or cocky with an undertone of anxiety.

Clients are good, said a partner once, slapping Patrick on the back.

But so many? said Patrick. He has only been in practice for three years—he tries to imagine the volume of clients after ten or twenty.

All these people, lost in their own private dramas—frantic about the results of wills, shell-shocked by the infidelities of their husbands, bankrupt and desperate to sell their houses. Strange that law—something that prides itself on subtlety, on intellectual nuance—so often involves people who are entangled in the most melodramatic and improbable circumstances. Sometimes he feels that their curdled emotions are beginning to rub off on him as well.

This is why I don't like opera, he said to Eduardo a few weeks ago.

This is why you don't like law, said Eduardo.

Don't be absurd, said Patrick.

Since then, though, this sticky little thought has been dogging him, turning up in unexpected places. Of course it *is* absurd. He enjoys law, he is good with clients—this is why he has so many. He is reassuring with them, he summons up a specific form of patience, a withholding of judgment about what they have done, a suspension of disapproval, no matter how well deserved. He thinks of excuses for their conduct, he tries to make them feel less embarrassed, he guides them through the various game boards of the legal process. He talks to them as if their idiocies were simply minor slips, as if threatening shareholders or inserting venomous paragraphs in marriage contracts were the kind of mistakes that anyone might make. He is even fond of them, some of them, anyway, the ones who seem to have a kind of erratic bravery, or at least, stoicism.

But some days he feels like shouting at them: *Why are you so foolish? Why are you so hapless? Why are you so cursed?*

These people aren't necessarily typical, said the partner.

I know, I know, said Patrick, and he does know. And this might not be their typical behaviour, either—some of them he is seeing at their worst. But these are the people he sees, day in, day out, and they seem to have an almost infinite capacity for poor judgment, for self-delusion. And as much as he tries to talk himself out of it, this stream of people is beginning to sour his view of humanity more generally.

Or maybe he is simply tired of making himself into the person they need.

But here is his next appointment, a man involved in buying a house and obtaining a *hypothec* on it. An assistant shows the man in, a man with a doughy face who looks unfinished. He stands on the balls of his feet, heavy, slightly pigeon-toed.

Patrick comes out from behind his desk, and shakes his hand.

'Good to see you again,' he says in French, although nothing could be further from the truth.

He gestures towards a chair, and they sit down to discuss the purchase.

'Unfortunately, I have bad news,' says Patrick. 'It looks like the sale is falling through. You may have to give up on this.'

'I want this house,' says the man.

'I know. But it may not be possible to get it,' says Patrick. He tries to infuse more regret into his voice as he outlines the problems in brief terms.

'I want this house,' says the man.

'I understand. I'm sure I would feel the same way if I was in your position. Have you taken a look at other houses in this area?'

'I want *this* house. Why can't you get it for me?'

Patrick explains again.

'I don't care,' says the man.

'You can't get this house,' says Patrick bluntly.

The man looks angry, and then his eyes narrow.

'You tell the seller that if I can't get this house, something might happen to it.'

'You're kidding,' says Patrick.

'No,' says the man.

'This house can't possibly be worth the risk of some kind of liability. Or criminal charges.'

'That's only if you get caught,' says the man triumphantly, as if he has scored a debating point.

Why are you so stubborn? Why are you so childish? Why are you so petty?

After he tells the man in several different ways not to do anything illegal, the man leaves, disgusted.

Not all his clients are like this man, Patrick reminds himself. Or at least, they rarely talk about breaking the law to his face. Usually law is something they want, quantities of it, as if it could be wrapped and delivered to them in packages. Or poured out of bottles, like very dry wines.

He understands some of this, or at least he thinks he does. They want confirmation of the stories they have developed about themselves, stories where they have cast themselves in the roles of the virtuous, the reasonable, the put-upon. At first this pliability with respect to the facts seemed odd to him. But now, sometimes he thinks they are right. He no longer begrudges them their chance to bend and colour their lives, or at least their descriptions of them. But he finds it difficult to understand their faith that the law will be on their side in this exercise.

They seem to feel this whether or not they have been disappointed in the past by law, whether the laws they have encountered turned out to be clumsy, or capricious. Over and over they fall into the same trap, hopeful, expectant, confident that the law will mend its ways, will do better this time.

He wonders if there is a legal doctrine for this. Or perhaps he should invent one. *The doctrine of lunatic hope. The doctrine of myopic hotheads.*

Something for everyone at the law store.

He glances at his watch—he is meeting a friend for lunch at a Vietnamese restaurant on rue de la Gauchetière. *You have to try the lime-chili scallops.*

The sky is a translucent pale grey, unfinished-looking, as if an artist had painted a wash on his canvas and then gone out for a drink. Patrick is walking only because the parking is hard to come by in this area—he finds walking tedious. Running, yes—he and Eduardo often run on Sundays, but walking—it should be obsolete by now. What is there to see? Pigeons, trees, cars, more pigeons, trees, cars. Closer to the restaurant, though, the street takes on a different appearance—open crates are set out on the sidewalk with white radishes, bunches of long beans, papayas. Stacked milk crates in front of stores hold twigs of lychee nuts, neon plastic sandals, pea shoots, paper umbrellas. Signs higher up offer acupuncture, electronics and medicinal herbs.

He feels a hand on his shoulder, and turns around to see his friend. The man has carefully shaped beard stubble covering a soft jawline, an argumentative mouth. He steers Patrick towards a small restaurant with Vietnamese characters on the window. 'It's better than it looks.'

Inside, a harassed-looking man in an apron gestures towards a table, and they sit down. The restaurant is decorated in red and gold, with Chinese lanterns along one side, and some Thai shadow puppets mounted on the wall. The tables have small red vases on them, with green bamboo stalks sprouting leaves. Patrick suspects the place has been through a number of incarnations.

The menu has the candour of literal translation, especially the soups—fatty flank soup, beef tendon and cartilage soup, and sliced beef stomach lining soup. They order the scallops.

While they wait, Patrick keeps up his usual flow of conversation, although the friend can hold his own, mostly talking about his cases in a choppy, barking way—one outrage, one miscarriage of justice after another.

Patrick wonders why he doesn't feel that way about his own cases—all this passion. But the friend is a criminal lawyer, so his clients go to jail. That might account for the difference in indignation levels.

'Tell me about arson,' he says, interrupting another tale of iniquity as the scallops arrive, surrounded by mint leaves and lime slices.

'What kind of arson?' says the friend promptly, ignoring his chopsticks and picking up a fork. 'There's arson with damage to property, arson with disregard for human life, arson for a fraudulent purpose, arson by the holder of beneficiary of an insurance policy, possession of incendiary materials.'

'Really?' says Patrick. 'I had no idea the world of arson was so comprehensive. Do you need intent?'

'Intent, recklessness, negligence—something, anyway. Depends on the Code section. Why?'

'Just curious. Did you see the news story about the market fire?'

'Yes,' says the friend, indignant again. 'We go to that market sometimes—Charlotte goes. She says it's one of the best places for cheese. Now we'll have to go somewhere else.'

'Well, that's certainly one perspective,' says Patrick. Although, he thinks, it might have been his own reaction, if he hadn't known someone involved. Or was more interested in cheese. One man's second-degree burns—another man's fleeting inconvenience.

'All kinds of arson jurisprudence,' says the friend. 'Regina v. Landry, if you want a recent case.' He is holding up a scallop on his fork critically, as if it were Landry, the unlucky accused.

'Maybe I'll take a look.'

Back in his office, he signs into a legal research database and tries *arson* and *negligence*. Eight hundred cases come up. He begins looking through them—Regina v. Eng, Harricharan v. the Queen, Regina v. Fewer—and then decides this is pointless. Either he should do a serious search or not at all. Too many cases. Too many Reginas.

A busy Queen, apparently. Not just a dowdy figurehead (or not only a dowdy figurehead), but someone prosecuting cases all over the Commonwealth, at least through her Crown attorneys. Interesting that Americans prosecute in the name of *The People.* Here, the job is left to the Queen—the people have better things to do. The majesty of the law, in a literal form.

Poor Queen, God save the Queen—not a popular person in this city. A stuffed figurehead, a remnant of history, wandering around the criminal courts with matching shoes and purse. The head of the Commonwealth, a collection of territories assembled by a voracious nation. Look, there is Canada, in the club with Australia, Saint Lucia, Belize, Jamaica, Tuvalu. The original fur hat providers, rubbing snowy shoulders with Antigua, with Samoa, with the Solomon Islands. What exotic, arbitrary company. But a little musty with age.

If only the wealth had turned out to be common. But as it is, not a popular concept here either. Ah, but he is a non-combatant—or worse still, a defector, someone who has fallen for the language of the other side. Seduced by how peppery it is, by the roughened textures that save it from over-prettiness, the rises and falls, the lilts and twangs. Although maybe these are things more likely to be perceived from outside a language—the shape and form more clearly seen by a non-native speaker.

He feels almost guilty about his attachment to French—almost, but not quite—as if the English community were some stray animal he was abandoning, now that its fortunes were in decline. But this is simplistic, absurd. What allegiance does he owe them? Or anyone else for that matter. He doesn't believe in tribalism. Don't we all live interstitially?

The refuge of the clever, says Geneviève.

Then the clever should be able to choose their company, including a set of live chess pieces on thrones. The possibility of going to Tuvalu and spotting the Queen's picture.

Look, we have the same Queen. You have the little finger on her left hand? We have her right knee.

The telephone buzzes. And here comes the next client, the next unreasonable optimist.

Forget language, he says to Geneviève. The real divide is between the hopeful and the cynical.

Not that she asked.

He thinks of the way she walks again, her bones gliding under her skin, and almost shivers.

I don't really sleep much any more, says Patrick.

I'm going to refer you, says the doctor.

Chapter Four

'What happens now?' says the cemetery director in French.

Eduardo looks around the table, an oak table with inlaid wood, a deep scratch along one side. Six other people are sitting around it, in this cool, damp room behind the cemetery chapel. The cemetery director—pebbly skin, blue slits for eyes—seems vague. This is his normal expression, as far as Eduardo can tell—he looks like the kind of man who often wonders whether leaving his first wife was a mistake.

The room is high and narrow, with an ornate plaster ceiling rosette, scrolled leaf cornices, a faint smell of mildew. A display shelf is stocked with pamphlets in French and English. *Caskets and Urns. Pourquoi cette mort? Services funéraires. Questions About Cremation. Cherchez un sens à cette perte.*

The mausoleum drawings are spread out in front of them, front views, side views, cut-aways. Around the table the faces are variously preoccupied, unsettled, guarded. Not a good sign—usually he can see at least a flicker of new respect at this point, sometimes even startled admiration.

'What happens now? At this point, we want you to be satisfied with the schematics,' says Eduardo. 'Then we'll do the design development, look at it in more detail, keep working on the practical problems.'

His ears still feel plugged.

From the explosion, said the emergency room nurse. From the explosion, said the emergency room resident. It will go away, said the attending physician.

When? When will it go away?

Soon, said the nurse. Soon, said the resident. Unless there's permanent damage, he added.

One of the building committee members—a pinch-faced man in an expensive suit—is picking up a drawing again, turning it around to examine it more closely.

All this, simply for bodies? But there is nothing simple about bodies. Collections of matter, tissue, fluids, bone cells in phosphorus—biochemically ingenious. Of course, Patrick had been talking about dead bodies—the detritus, something to be tossed aside when the last breath has expired, the last nerve impulse stilled. A shell, inhabited and then abandoned. And what remains afterwards? The remains. Lifeless bodies, the rubbish. Quick, bury them. Burn them. Encase them in stone. Or in this case, in metal-lined crypts with marble facings.

But surely corpses deserve this? After all the indignities, the pickling in formaldehyde, the makeup. Surely a ceremonial home is in order after that, somewhere to at least rot in peace, if not rest in peace.

Not very rigorous thinking, says Geneviève.

Not everything involves science, he says.

No?

The pinch-faced man tosses the drawing down on the table again.

'I think the building should look more church-like, more historical,' he says. 'I don't know why we need all this glass. We don't want people to feel they're on display when they're visiting the crypts.'

The junior next to Eduardo, a young man with pale skin, licorice-dark hair, stirs impatiently.

'What about adding a church part to it?' says another man with a baggy face. 'Like a bell tower. I think that could be very nice,'

'This is a modern building,' says the junior, whose name is Jeremy Boyer. 'The iconography should be appropriate to its time, not some historical cliché.'

Eduardo gives him a warning look.

'Maybe we can think about increasing the emphasis on things that churches provide for people, things like solace, reverence,' Eduardo says.

Is that smoke he smells? No, surely not.

'Well, why not make it look more church-like then?' says the pinch-faced man, aggressively. 'What's the problem? We're the ones paying for this.'

'We're trying to address issues of transparency, of time and space,' says Jeremy disgustedly.

There is a mutinous rumble around the table.

'Look, we're a small cemetery. And we have to sell these crypts to people,' says the man with the baggy face, annoyed.

The smell of smoke is getting stronger. Cigarette smoke? No, sootier than that.

'A fine building will help to sell the crypts,' says Eduardo hastily. 'We can make this something elegant that works commercially as well.'

'Don't worry,' says Jeremy, 'they'll be dying to buy these crypts.'

'We know what people want,' says the pinch-faced man, glaring at him. 'They're Catholics, they want a place that looks holy.'

'Of course,' says Eduardo. 'A sacred quality. We know exactly what you mean. Why don't we make some modifications and get back to you?'

'Yes,' says the director vaguely, before anyone else can speak. 'I think that would be best.'

There is a general movement, people standing up, shaking hands, gathering up notes. Jeremy is rolling up the drawings, putting them back into a cardboard tube, while the pinch-faced man says loudly: 'Are we tied to these people?'

On the way out, the director takes Eduardo aside, suddenly shrewd.

'It's not so much the design,' he says quietly, 'as the fire. They don't want to be tainted by association.'

'What?' says Eduardo.

'The market fire. What they're saying about arson.'

'What the hell does that have to do with this?' says Eduardo, bewildered.

'Is it true what they're saying about the owner?' says the director.

'Is what true?'

'That the owner burned it down for the insurance money,' says the director.

'I haven't heard that,' says Eduardo, startled. Although if the fire was arson, presumably the owner might be a logical suspect. No wonder the man isn't returning his calls. 'But why would it have anything to do with this?'

'They're saying that the money was going to pay for your renovations.'

'*What?*' says Eduardo, with a rush of anger this time. 'Exactly who is saying this?'

'It's just going around,' says the director, taken aback.

'*Who?*' says Eduardo. The smell of smoke is catching him at the back of the throat.

'It might have been Gauthier, Claude Gauthier. He's the director over at St. Pierre's Memorial. They're doing some building as well.'

'Look,' says Eduardo. 'I had nothing to do with that fire. I don't even have the commission yet.' He hears himself speaking with a strange, woolly-headed intensity.

'But you were there,' says the director, more boldly.

'Do you think if I had anything to do with it, I would be there when it happened? And get burned myself?' He holds up one of his gauze-wrapped hands.

The director shrugs. 'I'm just telling you what people are saying.'

'Listen,' says Eduardo, forcing himself to sound reasonable. 'This is ridiculous. I really didn't have anything to do with it. And I don't know the market owner, but it would surprise me if he did, either. Tell the others not to worry. But assure them that we understand their views on the design—that's the point of these meetings, to get feedback so that we create the building you want. I'll call you when the modifications are ready.'

Outside, the smell of smoke begins to dissipate—a summersweet bush is scenting the air instead, white flowers drooping in the hot breath of August. Rain begins pattering down in a desultory way, although not enough to cool down the pavement. He inhales deeply—as much to clear his lungs as to calm himself, so that he doesn't strangle Jeremy, who is leaning against the car, looking defensive and resentful. Not an unusual expression for him, Eduardo realizes. Was this recent? How long has he been at the firm? Three or four years? When did he acquire this sense of grievance?

Who cares? He isn't about to start coddling the juniors.

Jeremy says nothing until they are in the car, when he begins to speak in a rush.

'They don't appreciate what we're trying to do,' he says in English, twisting around to back the car out of the parking spot.

'What's the matter with you?' says Eduardo. 'Do you want to lose this project? You're going to have to learn how to handle clients better than that.'

'But they're philistines,' he says.

The rain is picking up, and he turns on the windshield wipers, smearing dirty water and a trapped leaf across the windshield.

'We don't know that. Or at least we don't know that yet. We have to bring them along, introduce them to the ideas gradually, make it a collaborative process. The last thing we want is for them to feel stupid, or that the project is too fancy or abstract.'

'But a bell tower?' he says stubbornly.

'Anything can be designed well. And you have to stop thinking that all that matters is the refinement and integrity of the structure. We have to create the kinds of meanings that touch people, that speak to them.'

'I know that,' he says, speeding up the windshield wipers, the trapped leaf whipping back and forth across the glass.

'Not well enough,' says Eduardo shortly.

Is he being unfair? If the fire was really the problem? But this kid does need to learn how to handle clients.

Christ, the fire. But it's ridiculous, he thinks. Who would believe it? The firm is reputable—more than reputable. Look at their awards—if anything, they are considered too arty, certainly not too venial. No, it's ridiculous. Who would believe it?

Except clearly the director, the other board members did believe it, or believed it sufficiently—enough to be wary, to mistrust him, to mistrust the design. Unless they didn't care whether it was true, but were concerned about their own reputations. Or perhaps both.

It wasn't only the arson rumour, though. He could see that they genuinely hadn't taken to the design. He is a little shocked by this himself—he has never had such an unenthusiastic reception before.

But he should have known—had he known?—that there was something missing in the design. It was too cerebral—technically elegant, but lacking some essential element, something that reaches into the psyche, that catches at the edge of possibility. That means when people look at the building, they don't say *how do you do this*? the way Patrick did. They say: *ah, now I see.*

Jeremy is still looking at him. He does have talent, he reminds himself. And inexperience can be cured.

'Think of Siza's apartment building in Berlin,' says Eduardo.

'*Bonjour Tristesse*?'

'Do you know how it got that name?'

'Because of the sadness of it,' says Jeremy automatically.

'And because someone climbed to the top of it, seven stories, and scrawled it across the facade,' says Eduardo. 'Illegally.'

The junior slows for a stoplight, looking puzzled.

'But maybe they were mocking it,' he says.

'It doesn't matter—you're missing the point.'

Jeremy looks resentful again.

When they get back to the office, Eduardo begins rummaging around in a file cabinet.

'I have some photos here somewhere,' he says, pushing aside file folders clumsily with his bandaged hands. 'Photos of a church in Évora, *Igreja de São Francisco*.'

He pulls out a large envelope and awkwardly shakes some photos out of it.

'Take a look,' he says.

The junior looks at them for a few seconds and then steps back quickly. The photographs are of an old chapel, a place with walls made of human bones—femurs, skulls with no noses, mouths full of mortar, no jaws.

'They're the bones of the monks, thousands of them,' says Eduardo.

'It's macabre,' says Jeremy. 'You're not suggesting we incorporate bones into the structure of the mausoleum?'

'Of course not. There's no question it's macabre. And primitive. But that's not the point,' he says.

The junior holds one of the photographs up and squints at it. 'Is this writing?'

'It's an inscription over the chapel door: *Nós ossos que aqui estamos pelos vossos esperamos*.'

He waits for Eduardo to go on.

'I don't speak Portuguese,' he says finally, when Eduardo says nothing.

'Think about it,' says Eduardo, gesturing towards the door.

His office is on rue Sherbrooke, an avenue that wanders across Montréal in a series of different disguises—a seminary, a Masonic temple, a park with a lagoon—drifting obligingly from English to French. The office has one airy studio for the interns and juniors—rows of desks with two or three computer monitors each—a boardroom painted in the colour they happen to be trying out that week, and some individual offices for himself and the two senior associates. The rain is still coming down, and the greyness outside makes his office seem warmer and more inviting. This is a trick, though—if he turns around, the office will become a place of business again, for all its aesthetic ambitions.

He rubs his hands across his eye sockets. The bandages are still hampering him, even though they are no longer boxing gloves, the tips of his fingers exposed. His eyes are stinging—surely they should be back to normal by now? It feels as if fine grains of pepper have been sprinkled under his eyelids. *Pepper in the eyes*—a faint echo of something comes to him.

Pimenta nos olhos. What is this?—a proverb, something from his childhood? His mother is a proverb hoarder, or maybe a proverb dispenser—she considers them the most infallible of truths, the explanation for every event, guidance for every contingency.

Signposts, he supposes—a kind of navigation. Not a bad one, either—particularly since she is someone who needs signposts desperately. If they give her something to hang on to, then he is all for them.

Pimenta nos olhos dos outros é refresco. The rest of the proverb surfaces. *Pepper in other people's eyes is a comfort.* An ugly thought.

His mother is not particularly mean-spirited, though. She has other problems—she is a damp, sluggish woman, growing larger every year, gently inflating. On her good days, she nurses a patch of garden, marigolds, bellflowers, some bush with large white flower balls. She eats a number of small meals throughout the day, always with a certain tenderness. On her bad days, she watches endless repeats of *Luso* Montréal or Portuguese telenovelas on the wide-screen television his father bought, sometimes mouthing the words.

Interesting, says Geneviève. I wonder what she thinks of them—they're so full of florid passion, all those defiant women. The furthest thing in the world from her.

True, his mother defers to almost everyone, to his father, to himself. His father takes this deference as his due, not cavalier about it, but not particularly appreciative, either. And she accepts this state of affairs without complaint, although lately he has noticed her talking about his father a little wistfully, as if he had floated down the St. Lawrence, instead of sitting there in the next room, reading the sports pages of *A Voz de Portugal*.

Was it being uprooted that did this to her? Even after so many years, she is now retreating, uncomplainingly, into her own bulk. And recently, she has begun exuding a faint smell, her breath, her skin, as if her internal organs were going off, fermenting a little.

His father doesn't seem to be affected in the same way—he is a straightforward exile, convinced of Portugal's superiority in all things, but resigned to living here for practical reasons. His mother seems to be suffering from a kind of chronic, low-grade mourning.

Another unwelcome thought.

But she doesn't have the consolation of work, the scaffolding it provides—or at least work besides housework, work that might give her another strand of selfhood, another sort of grouping. Unlike his father, whose work is so entirely practical, so plainly useful that it has protected him to some extent from feeling out of place. Any place. A mechanic is always in demand, a man with universal currency.

Like an architect? Not necessarily. Depending on the architect, he supposes. The paradox of being rooted in a place and transcending it. He rubs his eyes again. Maybe he should go home, take some painkillers—the skin on his hands is itching and sore at the same time—or take a nap. But Matty will be there, racing around. Matty. Christ. He should be glad that Geneviève has someone to distract her, especially since he himself isn't much company at the moment. But having Matty around is like being invaded by an excitable, spinning top.

Don't get too attached to him, he says. Don't forget he isn't your child.

How would I forget that? says Geneviève. Tell me how I would forget that.

The rain has stopped outside, but the sky is still overcast, lidded with heavy clouds. He can hear Jeremy and an intern arguing in the next room, and he feels his stomach twist. Two of the people who have bet their jobs, their career prospects, on a man with a headful of designs and an overextended line of credit.

We're not the only firm with financial problems, he reminds himself. It's almost endemic—this insane, all-or-nothing business. But for a second, he can see the stark outlines of failure, as if it were a computer-generated drawing, a cut-away view.

He tries to yawn, hoping to clear the feeling in his ears. But it isn't only his ears—or his eyes. He feels as if he has been sealed up in a glass compartment, isolated from everything, but at the same time, exposed.

Do something useful, he says to himself. Call the market owner again, find out what the hell is going on. Or call this other funeral director, track down where the insurance rumour is coming from. Maybe he can stop it at its source.

He picks up the telephone, stabbing at the numbers with a pencil in his bandaged hand.

'I'm not sure where I heard it,' Gauthier says warily.

'This is important,' says Eduardo in French. 'Crucial, really.'

He can almost hear the man thinking on the other end.

'I'm not out to go after anyone,' says Eduardo. 'I just need to get to the bottom of this, to clear it up quickly.'

The man is still hesitating.

'You're a businessman too, you know what something like this could do to a business.'

'It's just a rumour,' says the man finally. 'No-one's accusing you.'

'I understand,' says Eduardo.

'Steven Hegge,' says the man reluctantly. 'He's doing some work for us, an addition to the chapel. He was the one who mentioned it.'

Steven Hegge?

I don't believe it, he thinks as he puts down the telephone. A small time architect, a man with brown-grey hair—well-cut, but slightly long—who sweats a lot, a slack face with a fine-featured mouth. A striver, someone who puts forward an overblown view of himself, but was too ingratiating to carry it off. Someone who jumped on every faddish element, attaching them to mediocre designs, but who left clients surprisingly satisfied.

Is he capable of starting a rumour like this? It seems unlikely. If anything, he has tried to strike up conversations with Eduardo, to impress him with rehearsed phrases. He must have been merely passing the rumour on—he is certainly capable of being spiteful. But the thought of calling him is unappealing. Hegge would be only too delighted—and only too quick to imagine a debt to himself.

Jeremy pokes his head in the door.

'Busy?' he says.

'Come in,' says Eduardo, stifling his irritation.

He comes in with an intern, a young woman—Sandrine Nadeau, short, unnaturally red hair, black-framed glasses.

'The rotunda,' she says in French, part of a children's library they are designing. The junior keeps interrupting her, shaking his head, closing his eyes in feigned pain.

With each interruption, she becomes more animated, so that she is almost acting out the problem, recruiting her body, her arms. Jeremy becomes more focused instead, attempting to describe things with absolute accuracy, pin-pointing his thoughts to within a hair. Between the two of them, they are working themselves up, becoming louder, more emphatic.

His office is filled with their tense sparkiness. Sandrine is so emphatic now, he wouldn't be surprised if she leaped onto his desk and began declaiming from there. Then, suddenly, it seems to him they are only bickering.

'Stop it,' he says.

They stop, mid-sentence.

'Look,' he says, pulling over a piece of paper and trying to sketch something. But the bandages are still in the way, so he tries to explain it instead.

They seem unsatisfied, looking at him expectantly. But who was right?

'Go away,' he says.

They file out, Sandrine a little abashed, Jeremy sulkily, almost colliding with Patrick, who is at the door, two cups of coffee in his hands.

'The famous Cabral tact,' Patrick says in English, putting down the coffee, and taking off his wet raincoat and shaking it.

'Nobody knows how to draw any more,' says Eduardo moodily. 'They've been ruined by software.'

'Modern youth,' says Patrick with an exaggerated sigh.

Eduardo ignores this. Talking is suddenly making him feel flat and tired.

He envies Patrick, still in his thirties. There is something about him, a certain buoyancy. Less than before—his divorce has seen to that, but he still has a sense of expectation about him. More than a youthful body, more than a smooth face, Eduardo wants that spatial feeling of time again, that feeling of horizon. Instead, he often has a sense of foreshortening—a leak in the quantity of available future.

'Well, I guess that was it for small talk. Let's see this contract for the gallery,' says Patrick.

Eduardo fumbles for a minute through the papers on his desk, and produces a manila envelope.

'I want to ask you something, too.'

'Fire away,' says Patrick, taking the envelope, and settling down with his coffee. 'The best notary in the province is at your disposal.'

'Why are people such idiots?' says Eduardo abruptly. 'They'll believe anything, they'll pass on anything. No-one seems to have any critical faculties left, no-one seems to think twice about repeating the most ridiculous garbage.'

'Such as?' says Patrick calmly.

'There's a rumour going around that the market owner started the fire for the insurance money.'

'Not that I want to join the ranks of idiots,' says Patrick, 'but I'm not sure why that's so ridiculous. They've classified the fire as arson, and people do set fires to collect insurance money from time to time.'

'The rumour is that I was in on it somehow, that the money was to pay for the renovations.' Even saying this makes him feel as if a black balloon is swelling in his head.

'Really?' says Patrick. 'Then they *are* idiots. If only they knew what a stiff-necked, hew-to-the-straight-and-narrow, pain-in-the-ass client you are.'

'Isn't there something I can do about it? Legally, I mean. To stop it. Isn't it libel or something?'

'Slander,' says Patrick automatically. 'Or defamation, really. But you have to know who's doing it, you need someone to sue.'

'How am I supposed to know who started it? All I know is that a couple of people have passed it along.'

'Hard to sue them, if that's all they're doing—you'd have to show that they knew it was false. Or that they should reasonably have known it was false. You're in the wrong province—Québec law is more lenient on this. Have you talked to the market owner yet? He must have some of the same concerns. Or maybe he can at least put it out there that you don't even have the commission yet.'

'But that won't help if we get the commission later,' says Eduardo. 'And it might make him less likely to give it to us, if it looks like it will add strength to the rumour. No, I don't even want to raise it with him.'

'Have you talked to him at all?' says Patrick.

'No, I've left a couple of messages.'

'Keep trying,' says Patrick. 'Talk to him and find out what's happening at his end. Or if he knows where the rumour is coming from.'

'And then we can sue?'

'Maybe, but you don't want to rush into anything. I can talk to a litigation lawyer for you, but the received wisdom with these things is that a defamation suit can keep something alive in the public eye, when it might have blown over otherwise.'

'For Christ's sake,' says Eduardo viciously. 'You're damned if you do, and damned if you don't.'

'What a surprise,' says Patrick.

I give up, says Jeremy. What *does* it mean?

Nós ossos que aqui estamos pelos vossos esperamos, says Eduardo. *We bones that are here, for your bones we wait.*

'Finally,' Eduardo says, holding the telephone awkwardly. Then he realizes that this might be impolitic and adds: 'I've been waiting to hear from you.'

Geneviève looks up from where she is settling Matty down in a stuffed chair in the living room.

'Yes,' says the market owner in a brisk way. 'Well, I'm calling to let you know that an insurance investigator will be around to see you.'

'Is that normal?'

'I don't know,' says the owner coolly. 'I've never had a building burn down before.'

'But you've heard the rumours?' says Eduardo.

'I've heard something.'

'Where are they coming from?'

'How should I know?'

'Look,' says Eduardo, 'If we knew where they were coming from, we might be able to stop them, take legal action.'

'I've been told that might not be a good idea,' says the owner, but he sounds less stiff. 'Besides, they're already out there.'

Eduardo is startled by the owner's resignation. Doesn't he care? What about his own reputation? Won't this affect his business as well?

'I need to know,' says Eduardo flatly.

Silence on the other end.

'I heard it from an architect called Hegge,' says the owner cautiously.

Hegge again. He certainly has been busy, Eduardo thinks grimly.

'I don't think he's really capable of starting a rumour like this,' he says to the owner.

'So he was only passing it on?' says the owner.

'Maybe I'll ask him where he got it from,' says Eduardo. 'Anyway, this insurance investigator—of course I'll talk to him, but there's not much I can say. And I'm leaving for Portugal in a few weeks, so tell him to call soon—I don't want to hold this thing up.'

He is on the verge of suggesting another meeting with the owner, to carry on with the discussion about the commission, but something about the owner's tone, his wariness, stops him. Maybe he should wait until after the insurance investigation. And he doesn't want to seem too eager, too much in need of the project. Particularly since he is.

As he puts down the telephone, he wonders whether the investigation will help. Perhaps it will be a vindication of sorts, but it might also add fuel to the rumours, particularly if the outcome is ambiguous. He refuses to consider the worst possibility—that the market owner was actually involved. If he takes this thought seriously, he will be as bad as the others, indulging in the same small-minded willingness to believe the worst. Is this some version of *schadenfreude*? Although maybe he should be careful about putting the market owner and himself in the same category, the category of the unfairly tarred. Just in case.

Why was the market owner talking to Hegge, anyway?

Suddenly he smells smoke again.

'Is there something on the stove?' he says to Geneviève.

She shakes her head. She is reading to Matty from a Tintin book, showing him the pictures. He is leaning over her lap unselfconsciously, engrossed, his elbow on her upper leg, a half-eaten croissant forgotten in his hand.

Eduardo almost envies the boy this complete absorption, this heedlessness of everything else. A brain not yet cluttered with the accumulated sediment of living, someone who can be wholly concentrated, every part of his body engaged—even if only for short periods. He must have been like that himself as a young child—he tries to remember how it felt, to recall that specific state of mind, but it remains elusive, perhaps because he was not conscious of it at the time. Only a place flickers across his mind—a kitchen with salted cod soaking in water on the table, a block of quince *marmalada* and a fly buzzing over a used knife. The smell of frying lard, the sound of rain drumming on a tile roof, a streetcar rattling by in the distance.

Our Lady of Fatima presides over the room from a corner, a small ivorite figurine tranquilly surveying her domain. Toys are scattered around the floor, a tin car, a plastic rooster, a picture book, a jumping jack.

No, the jumping jack is missing.

These things can't disappear by themselves, said his mother placidly.

But to Eduardo, the child, this seemed to be precisely what happened—his possessions were constantly moving, creeping around like snails.

Was it something to do with their apartment? Of course not, the apartment was nothing out of the ordinary, a whitewashed building with double doors at the bottom, red carnations spilling out of the window boxes. And the street in Alfama district, an old narrow street where he spent most of his time, where even now he could describe a patch of missing paving stones, a metal pipe valve set into the wall. The pipe valve had raised letters on it—*água*—letters he would trace with his finger when he was bored. Or he would pick at the white paint flaking off the wall in the sun.

None of this explains the slipperiness of his possessions, though. Arriving in Montréal, overwhelmed by upheaval, even then his things continued to disappear and reappear. So this was proof—the problem was him, not their apartment, an innocent flat—or as innocent as a childhood home can be.

Inconvenient? Yes, but he is used to losing things after all this time, even resigned to it. He accepts that the effort of having to look for them is unavoidable—some personal kink in the rules of physics.

If he were being honest, he would have to admit that he often has a sense of over-accumulation. The idea of losing things helps to relieve some pressure, although this method—if it can be called a method—is a little erratic.

There are other ways of weeding things out, says Geneviève, puzzled by the disappearance rate of his possessions, their possessions.

But is it more difficult to keep things, or to lose them?

Saint Antoine, Geneviève says in a singsong voice, half-laughing. *Saint Antoine de Padou. Toi qui as le nez rendu partout, prie pour nous.* Patron saint of lost things. You whose nose is everywhere, pray for us.

He wonders what they look like now, his old street, the house. He hasn't seen them for years, since his last trip home. Probably much the same. Then why does the thought of this seem exhausting? Why does everything seem exhausting at the moment?

'*Dis bonne nuit à Eduardo*,' says Geneviève, smiling, carrying Matty off to bed—their study has been turned into a makeshift bedroom.

'*Bonne nuit*,' says Matty, waving both arms at him.

Eduardo watches Geneviève's retreating back. Don't forget he's not yours.

She glances back over her shoulder. Don't be such a killjoy.

'What did he say?' she says a few minutes later, coming back into the room without Matty.

'The insurance company is investigating,' he says.

'That's good, isn't it?'

'I guess so. Unless it really was arson.'

'You think the market owner set it?' she says, dropping down beside him on the sofa.

'I don't know, it seems unlikely. But I don't know the man, I don't know what to think.'

'At least it should clear you,' she says, putting a thin arm around his shoulders, rubbing them gently.

'Christ, I hope so. If the investigator has any brains. And if the damage isn't already done. I keep running into people who have heard the rumours. I can even spot them now—there's a particular look they get, a kind of wariness. As if they're assessing me while they're talking to me—is he capable of this or that? And they're more stand-offish, as if they're hedging their bets, just in case.'

'What did he say about the commission?'

'I didn't ask him,' he says. 'I'm afraid he might say no just to dispel the rumours. I'll follow up with him as soon as the investigation is over.'

They sit there for a minute, then he sighs—a long, ragged sigh.

The nurse unwinds the dressings from his hands—the skin on the back of them is yellow, grey, brown. Then she begins cutting off dead skin—in some places the skin lifts off in pieces. Underneath are wet, red patches. He tries to look away, but his eyes are drawn to the repulsiveness of it.

'Looking good,' says the nurse in French, snipping away at the dead skin.

Maybe it looks better in French.

'How long will it be before they're better?' he says.

'Another couple of weeks, and you should be fine without the dressings,' she says.

No, he thinks. How long will it be before they look like they did before? But what if the answer is *never*?

This is weakness, this is vanity. Hands are tools, the essence of practicality. As long as they work, who cares what his hands look like?

He does. What if his own hands are turning into strangers now?

I know it like the back of my hand. Maybe not any more.

Amor, fogo, e tosse, A seu dono descobre, says Eduardo's mother. Love, smoke and a cough are hard to hide.

An old wives' tale, says Geneviève.

What is the new wives' tale? says his mother.

Chapter Five

This is a riddle. This is a test, says Geneviève. A scroll, a neck, a bridge, four strings. What is it?

Je sais pas, says Matty.

That's right, she says, as if he had guessed. There's no fooling you.

No getting around him, either. She had assumed that Eduardo could look after him while she went to a rehearsal, but he had gone to work.

Do you really have to go in on a Saturday? she said crossly. He looked at her blankly.

Luc was away for the weekend and she had gone through everyone else in the family except Pauline, who was too far away. They were busy, coming down with something or otherwise unable to take a small boy who couldn't sit still. In desperation, she thought of Patrick—maybe he had Chloe this weekend?

I know this is an imposition, she said hesitantly.

Sure, bring him over, he said. He can help entertain Chloe—I think she has crush on him.

Lucky Matty, she said lightly.

Lucky Chloe, said Patrick.

Now she is rushing, she is late, she is looking for a street number on the east side of Mont Royal. The houses here are large and surrounded by well-groomed shrubs, artfully placed, everything here seems more substantial, sleeker—the sidewalks made of heavier concrete, the cars of denser metal, the trees lusher. Some of the shrubs have been trimmed into a set of vertical ruffs, as if a poodle groomer had been doing the gardening.

She shifts her grip on the viola—at least she doesn't play the cello. She thinks of the cellist, hauling his instrument around in its black case—like a mammal of some kind, a sea lion that has strayed onto dry land.

I don't see the point of this, the first violin had said. He is an acerbic man, good-looking, but in a slightly ratty way. Why do we need to rehearse at the house itself? The acoustics will be different for the concert anyway, when the place is filled with people.

But we need the rehearsal, and we might as well get the feel of the place, said the second violin diplomatically. She has recently taken on the task of handling the first violin, much to the relief of Geneviève and the cellist, a nervous man with a paunch and large eyes.

Ah, here it is—one of the bigger houses, with a chokecherry tree on the front lawn. She rings the bell, knowing that the second violin and the cellist will be there already, but hoping that the first violin will be late, as he often is. Except he is the one who opens the door, frowning at her—self-righteous as only the occasionally virtuous can be.

The house is draughty, but handsome, with white walls and marble floors. Down a step there is a living room, large enough for a salon.

'Sorry, sorry, everyone. What a place,' she says in French.

There is a chorus of agreement from the other three musicians. Chairs, music stands are adjusted, scores opened. She takes her viola out of the case, tightens her bow.

The owner of the house comes into the room.

'Normand Coté,' he says, holding out his hand. He has a long, ruddy face, a small mouth, white hair.

'This is a lovely place,' says Geneviève, after she introduces herself. 'Thanks for having us here.'

'My pleasure entirely. Do you mind if I listen to you rehearse? I'll stay out of your way.'

'No problem,' says the first violin, not to be outdone in politesse. The cellist looks mournful.

They tune up for a few minutes, and then they are into the first piece, a Shostakovich quartet.

The movement has a turbulent opening, with swooping high notes. Then a pizzicato part, an intelligent flight of sounds. The next movement is slower, forest-like, with a high violin. After this, there is an allegro, beginning abruptly, winding up to a taut peak, and finally dropping.

The cellist is a hair's breadth slow, and they try it again. Geneviève stumbles in a difficult passage, and they laugh, except for the first violin. Then they remember that the house owner is listening.

'*Tempo rubato,*' says the first violin. He wipes his forehead with the white cloth he uses on his chin rest, a gesture that Geneviève has always suspected was more theatrical than practical.

He does perspire a lot, says the second violin calmly.

But even in rehearsal, as they move through the Shostakovich, some Haydn, Grieg and Brahms, the instruments combine into complex, brilliant sounds. With the music flowing through her—from her—around her—Geneviève feels a familiar exhilaration, an almost physical gladness thrumming through her body. As if her nervous system had been replaced with warm, pure sound. Blotting out any thoughts of Eduardo, of Matty, of Patrick—any other thoughts at all.

When the music stops, the owner is sitting with his eyes closed, his face rapt and serious. He opens his eyes.

'I'd love to have these salons on a more regular basis,' he says. Then he becomes host-like again, offering them espresso, drinks, a tour of the house. They refuse the drinks, and take the tour instead. He leads them through a series of distilled, graceful rooms, pointing out a carpet, a painting. There is nothing boastful in this, nothing in his voice that sounds like pride of ownership. Instead, he seems thoughtful, respectful about these things. Although he has a faint air of expectation, she thinks. Almost as if he assumes the musicians will find these objects familiar, as if they were distant cousins themselves.

'Where did you learn to play?' he says politely to her in French, as they go up another staircase.

'I started out at a community centre near us,' she says. 'They had music lessons for neighbourhood children.'

He is duly impressed by this, by the whiff of triumph over adversity.

She refrains from saying that her mother made them take everything, that she seized on any kind of schooling or lessons available, for the girls as well as the boys.

You think they're going to be bank presidents? said her father. They've had enough school—they should have jobs by now.

I don't want my girls cleaning houses for *les Anglais*, said her mother pointedly.

There are worse things, said her father.

True, thinks Geneviève. Like being someone else's retribution.

On the third-floor landing, Normand picks up a cracked ceramic bowl sitting on an antique washstand.

'Can I hold it?' she says. He hesitates for a second, and then passes it to her. She turns it over, and traces the crack with her fingers.

'Wabi-sabi,' he says.

'The horseradish?' says the first violin.

'No, that's wasabi,' he says. 'Wabi-sabi is an aesthetic—a way of looking at things. About the beauty of things that are flawed. A cherishing of imperfection, of transience.'

Geneviève feels as if she has just discovered her side of an argument. You see? she wants to say to Eduardo. But see what?

Then they are in the last room on the third floor, a series of mobiles hanging from the ceiling, or poised on stands. Most of them are abstract—eccentric shapes, unusual colours. Their parts are delicately counterbalanced in ways that seem improbable. Some of them are in staggered formations, others extend outwards in a series of arcs.

Normand begins walking around the room, flicking them, so they are all in motion at the same time. Each one has a different movement, oscillating, bobbing, seesawing, gyrating.

'Wonderful,' says the second violin. Geneviève nods.

'I like to collect them. But most of them aren't particularly valuable,' says Normand.

'No?' says Geneviève.

He certainly is a handful, says Patrick, handing over Matty.

Merci beaucoup, beaucoup, she says. You're a lifesaver. A people-saver. Saint Patrick.

One of my manly attributes, he says modestly. Saint Patrick, though—there's an idea for a career change. If only I were Catholic. Or, come to think of it, Irish.

Not Irish, she says. It's been done.

What would you prefer, then? I'm freelancing for a cultural identity at the moment. Italian? Russian? Swedish?

He looks suddenly weary, though, and she is seized with the urge to run her hand along the side of his face, along his shoulder, down his arm, to take his hand and hold it to her chest.

I'm late again, she says hastily, pulling Matty away.

What was *that*? she says to herself, startled.

'You would have loved this house,' she says later to Eduardo.

'Why?' says Eduardo.

Would he? How can she describe it?

'I don't know, exactly—there was something unusual about it,' she says.

He waits for her to go on.

'It wasn't just beautiful—it *was* beautiful, but it was a kind of muted, worn-out beauty,' she says. 'Subtle. Or a bit askew or something.'

She tells him about the idea of wabi-sabi.

'Oh,' he says, in a dispirited way. 'I guess I'll see it at the concert.'

He would never have let that go before. Would he?

Matty. *He certainly is a handful.* More like an armful, an earful, a headful. A four year old shouldn't need watching every minute, but this one does. Although there is something about his rambunctiousness, his odd relationship to the physical world, the way he is constantly astonished at the results of his own misguided forays, that Geneviève finds endearing, that makes her laugh.

Yes, says Marie-Thérèse, who calls every night to talk to him. He's a live wire, all right. But she says this indulgently, proudly.

At the moment, he is temporarily transfixed, watching the city slide by outside the bus. They are on their way to the doctor, the last place she wants to take him. Who needs a child running around a waiting room full of childless people? But she has already dropped Matty off twice this week with Luc and his girlfriend to teach classes, she has a meeting coming up, and they will have to take him again when she and Eduardo leave for Portugal in couple of weeks. And Patrick? The thought is tempting, but he wouldn't have Chloe on a workday.

So here they are, in the waiting room of the doctor—a taciturn fertility specialist. Although his specialty seems to be providing the least amount of information possible, she said once to Marie-Thérèse. She has given up asking him questions because these turned into fencing matches in which he became increasingly noncommittal.

Matty is tearing pages out of a magazine, and the woman sitting next to them is watching him intently. Geneviève glances at her—the yearning in her face is unsettlingly naked. Then she sees Geneviève looking at her, and her face closes up.

Don't worry, Geneviève says silently. Don't be misled. I'm one of you, a fellow refugee, seeking asylum from the country of infertility.

Why is she the only one in her family with this problem? Her brothers and sisters seem to be able to reproduce on demand—and sometimes even without demand. All her nieces and nephews—when they are together, she feels as if she is surrounded by small people with half-familiar smiles, hair, gestures—pieces of her brothers and sisters tacked on to other faces. Odd, these little part-copies—genetic ingredients, combined and shaken up in a biological cocktail with such haphazard results. More than that, they seem to have been born superb mimics, taking on a facial expression, a posture, a way of walking as easily as breathing. Anyway, no need to worry about a *crise démographique* here, the lack of little Francophones—her family has been churning them out.

Although maybe not in the numbers of her mother's generation, when they had eight, nine, twelve children. Of course, they were performing a sacred duty, producing little Catholics. Politics, patriotism, religion—all reduced to a matter of eggs and sperm. She imagines a chorus line of haploid egg cells in a row—some with fleurs-de-lys on their chests, others with crucifixes. All kicking in time to the *cancan.*

Although her mother only had six children—almost an underachiever. She is short, stumpy, a pointed chin, a print apron wrapper over a print dress. Eagerly kind, longing to share her roster of saints for every malady. Too busy with her children to learn English, not even how to say *hello* or *yes.* No need for it, anyway—she could rely on her *Anglais* husband.

Not a timid woman—she is often opinionated—but she panics when the voice on the telephone is English. Even after they coached her to say *'no speak English'*—three little words—she would freeze, holding the telephone receiver away from her body, as though the English words at the other end might leap out of it and attack her.

Is this why Geneviève still feels wary of English herself? Not just a matter of allegiance, but something more visceral. Or is it because she feels flatter, more diffused, when she speaks it?

Her mother is more certain of herself in other ways, though, unlike her father, who seemed to become increasingly transparent with each job lost, with each failed job interview, as if they were draining him of some quality of materiality. Although he is dignified about this, not unlike a kind of sentence he was required to carry out.

He had a ritual for his job interviews—hand-starching the shirt he would wear. No-one else was allowed to do this, and it had to be boiled starch, not the spray variety. She can remember the smell of it—he added cold water to the raw starch first, and then let her stir it until it dissolved. After that he would add hot water from the kettle, and when it was thick and white, he would dip his collars and cuffs into it. He would iron them himself when they were dry, although he was content to leave the other washing to her mother.

Did this work? Did it help him get jobs? Of course not.

One of the family losers, says Raymond, her oldest brother, a burly man with an antacid habit. Where do they come from? Is it some bad gene? Maybe we should be watching for the symptoms in ourselves.

Although he was imported into the family, says Jocelyne. Maybe we attract losers as well.

Think of it as an aptitude, says Luc. A gift for misfortune.

This is stunningly unscientific, says Geneviève.

'I'm putting you on a new medication,' the doctor says in English, when she is finally called in to his office. He looks at Matty incuriously, and runs through the possible side effects—tremors, ulcers, bone-thinning, delirium, mania. So much to look forward to. Heightened libido—he says this with an entirely straight face. Perfect, now that

Eduardo doesn't have the use of his hands. Or apparently, certain parts of his brain.

'Do you think it will make any difference, going away?' she says. Different water, different air. Infertility—so enigmatic, at least their infertility, that everything is a potential factor, an agent for or against.

'Hard to say,' he says.

Of course—the emperor of non-information. Parting with a fact might strain something in his system.

'Let's go,' says Matty in French, tugging at her hand.

When they get home, she sets up some paints for him, a jar of water, paper, but he has already disappeared.

'Where are you?' she says in a singsong voice, pretending to look behind chairs, in cupboards.

'Are you under the bed?' she says, crouching down to look. She can hear him giggling.

'Are you in the closet?' she says, opening the closet door. He falls out, bursting with glee at his own cleverness.

She watches him paint for a while, splatters of colour running together on his damp paper, while the water in the jar begins to turn muddy. He starts wriggling again, and she brings a cloth over to wipe him off. He lets her do it, holding his half-curled hands out patiently, protesting only when she wipes his face. Then she takes him into the bedroom to jump on the bed.

Five minutes, she thinks, leaving him there. This will give her at least five minutes to herself, to confirm the arrangements with another professor to cover her classes while she is away. Five minutes before he gets tired of jumping, before the bedclothes start slipping off the bed, before the bed starts migrating across the floor. But only a couple of minutes later, she realizes that the bed isn't creaking any more.

'Where are you?' she singsongs again, the same routine, opening the cupboards, looking under the bed, opening the closet suddenly. But he isn't there.

'Where *are* you?' she says. Then she notices the balcony door is open.

There he is, arms hanging over the wrought iron railing, watching something below. A rush of fear, and she has him, pulling him away.

'*Non, non, non, non, non,*' she says sternly, bringing him back inside, closing the door with her foot. How did he get out there? She must have left the door open—she is aghast. Such a tiny thing. Such a terrifying possibility. They will have to keep the door locked.

She runs the water for his bath—more for entertainment purposes than anything else—and gives him a colander, a ladle, a mixing bowl, some coloured measuring cups. He stands up in the bath, enraptured, scooping up water in the colander and watching it turn into rain on his legs. His small body is marked in sun lines, brown legs and arms, a white torso. His penis is a small rubber tube that bobs when he stamps in the water, when he laughs. In a few minutes, though, he is trying to clamber out of the bathtub. She wraps him in a towel, and lifts him out. He is quiet for a few seconds in her arms.

'What am I going to do with you?' she says, holding him close, feeling his wet hair under her chin, against her neck.

He is in bed by the time Eduardo arrives home that night. She suspects that he has stayed at the office later than usual for this very reason. Will he be like this with their own child—if there is one?

No, this is different. He doesn't know Matty, Matty has been foisted upon him. And he seems more self-conscious about Matty than she is, wary of the boy being a stand-in, a substitute child. Concerned that having him there underlines their own lack of success, makes them look foolish, or even pathetic. Of course he would feel differently about his own child.

(Although Patrick was willing to take him. But that was different. How different? Somehow different.) And Eduardo is still so out of it, so unlike himself. Give him time.

The market fire, still wreaking havoc. The after-effects drifting into her own life, like grey ash deposits. I wasn't even there, she wants to protest, as if this made it more unfair. How much time?

She is sitting in a committee meeting the next day, listening to a discussion about curriculum reform. Does the curriculum need reforming? Maybe, but this isn't the way to do it. The discussion is dominated by the chair of the committee, a pretentious geneticist who bullies the junior faculty members, and fawns over the senior ones.

Surely he should try to be more subtle, she thinks, as he snaps at a young bryologist who voiced an opinion. His goal seems to be to reduce any issue to the smallest point possible, and then pronounce it resolved, congratulating himself on another one of his achievements. She feels a brief flash of gratitude for her department head—ambitious, yes, odd, yes, but at least he isn't venomous like this man. This man is—what? Fungi classification? Call him *amanita ceciliae*—snakeskin grisette. Or one of the stinkhorns.

And the others? What fungus are they? The white-haired, kindly neurobiologist who is intervening now on behalf of the bryologist—he must be *hericium erinaceus*—lion's mane.

The jumpy, middle-aged athlete is *trametes versicolor*—turkey tail. The mild plant biochemist with a kind of bristly intelligence—she would be *tremella foliacea*—leafy brain. What about the red-faced molecular biologist who cultivates graduate students as if they might burst into bloom? (Indeed, she sometimes wonders if he repots them from time to time, tucking his own patented intellectual earth and compost mix around them.) *Boletus rhodoxanthus*—ruddy bolete?

She tries to think of an excuse to leave early, but nothing comes to mind that would be good enough. Then pay attention, she says to herself, but this doesn't work either. Instead, she goes back over the rehearsal in the elegant house—the wabi-sabi house—over the passage where she had stumbled, hearing the notes in her mind, her fingers twitching.

How right brain, said a British mycologist at a conference, when she let it slip that she played the viola as well. He sounded patronizing, as if she couldn't possibly be either a good mycologist or a good musician.

I don't play very well, she had said, almost defensively. But she does play well—although this is something she only admits to herself. So why say this? To reassure him? But why should she have to be defensive about this?

How wrong brain, she wishes she had said instead. You call yourself a scientist?

You're too self-deprecating, the department head said to her once. We can't afford it.

Which *we*? she wanted to say. We, the Department of Biology? We, the University? We, *les Québécois*? And how much self-deprecation is too much? When did self-promotion become a requirement for everything? Let's bring back humbleness—much more appealing. Also likely to be more realistic.

But she knows there is something to what he says, some annoying bit of insight. One of those insights that leads almost nowhere in practical terms, that simply lies there, inert. Or maybe it leads almost everywhere—equally unworkable.

Self-promotion is overrated, says Patrick. Almost by definition.

A twitch of warmth runs through her body. Go away, she says.

Professor Turkey Tail is tapping a foot restlessly now. He has a physicality about him, as if his own body is always present, always in the foreground of his mind. Maybe they could all start tapping their feet, like prisoners clanking their forks in unison, until the meeting is adjourned. But the committee chair might not even notice for a while, he is so caught up in his own show.

When they are finally released from the meeting, she stops for a minute to talk to the bryologist. His research focus is on rock mosses—or is it feather mosses? Ah, mosses—not as delightful, not as gorgeous, not as charming as fungi, of course—but still interesting. At the moment, though, the act of talking to him is the important thing—an implicit gesture of support. She looks around to see if the poisonous committee chair is watching, and feeling mischievous, waves at him across the room.

He raises his hand uncertainly, and she turns back to the bryologist, who is aggrieved and only too happy to tell her all about it. She listens for a few minutes, and then manages to extricate herself with the excuse that she has to get back to her research—which is, of course, quite true.

In fact, this is all the more important now, she thinks, back in her office. What if Eduardo's reputation is so damaged that the firm really does go under? What if she becomes their main source of income? Maybe she shouldn't be going out of her way to antagonize other faculty members. What if Eduardo doesn't return from the land of remoteness and distraction?

But this is an Eduardo-type thought, this is Eduardo rubbing off on her. If she starts thinking like him, they really will be lost.

Tempo rubato, says Geneviève, thinking of the first violin. *Stolen time*. Borrowed from one line of music and added to another.

They are walking along St. Urbain, and the evening air is mild and gritty. As they draw closer to the church, Geneviève can see strings of coloured lights in front, creating a glittering net in the darkened sky. Underneath this is a makeshift courtyard with rows of pennants hanging on each side. The pennants are all the same, white with red Maltese crosses. The place is crowded, and the air is filled with the smell of garlicky meat and bread. A band is playing in fits and starts.

'*Assunção,*' Eduardo says to her, waving his arm dismissively, as if he were presenting all this to her and Matty. Yes, the Feast of the Assumption—the Portuguese version—has staked a claim to this small part of Montréal, at least for tonight.

They join a long line, curling around the courtyard several times. People are shouting greetings at Eduardo, clapping his shoulder, asking about his hands. Over to one side, she can see his father talking with a group of people, his wife firmly in his grip.

'Hold our place,' Eduardo says to her. Matty is jumping up and down, leaking excitement.

Eduardo disappears for a few minutes, then returns with red wine in plastic cups, some Orangina for Matty.

'From a barrel,' he says, passing her the wine. 'They make it in the church basement.'

'What an excellent use of a church basement,' she says, taking the glasses from him. 'What are we lining up for?'

'*Bolos de caco*,' he says, pointing to a shed where flat loaves are baking over a wood oven.

As they get nearer, she can see a small assembly line of people kneading dough, or pressing it into pancake shapes. A woman in a black kerchief is checking the loaves baking on the top of the stove, her forearms dusted with flour, a trickle of sweat running down her forehead. Five dollars each. They buy a loaf and Geneviève shuffles the hot bread between her hands, pulling off pieces for the others. She is hungry, and the bread is sweet and chewy.

Men are still slapping Eduardo on the shoulder blades, tapping his chest with the backs of their hands, some of them with children in their arms. She can see them asking about his bandages, his explanation, over and over. She can see them asking about Matty, his explanation, over and over. No, not a son, a nephew, her nephew. She nods and smiles, hardly hearing them above the noise of the boisterous crowd, and busy keeping an eye on her not-son.

Even here, though, Eduardo seems more subdued than usual. He seems almost—what? Almost nervous of the crowd, as if it might engulf him, as if the people might turn on him. Although he does introduce her to a couple of them.

'He used to live near me in Lisbon,' he says to her, after one of these introductions. 'When we were kids. His father sold chemicals.'

'Chemicals? Like a pharmacy?'

'No, like blue copper crystals for bleaching, lemon oil for furniture, that kind of thing.'

'Is everyone here from Lisbon?'

'No, not at all. Most of them are from the Azores, some from Brazil.'

His father waves them towards him—close to some fire pits made from metal drums. Eduardo shakes his head, but his father only waves more insistently. They go over, Eduardo standing as far back as he can, holding up his hands ruefully when they are handed skewers of meat, sprinkled with coarse salt and bay leaves. His father takes the skewers and stands over one of the pits, rotating them until the meat is grilled, while Eduardo watches from a distance, and Geneviève chases Matty.

When the meat is done, Eduardo's father pushes it off the skewer onto paper plates for them. They wander through the crowd, Geneviève carrying the plates, and sit down at a long table with other people—the usual greetings for Eduardo, the same questions.

She begins sawing away at the meat, which is fragrant but tough.

'Hold on to your teeth,' says a young woman in Portuguese beside her. She is putting on dark lipstick, rolling her lips together. When she sees Geneviève's uncomprehending face, she repeats it in French.

The rest of the people at the table are still speaking Portuguese, interrupted with bursts of laughter. Geneviève has tried to pick up some words before, but this is too fast for her. The woman with the lipstick begins to translate.

'They're talking about how they used to steal kumquats when they were children,' she says, pointing to Eduardo and the other man from Lisbon. 'Then they lined up on a wall to see who could spit the stones the farthest.'

She listens again for a minute.

'Now they're talking about how kumquat stones were the best for spitting. Big and slippery.'

She rolls her eyes, and she and Geneviève laugh.

'Come on,' says the young woman. 'Let's walk around. We'll let them talk about the good old days, when boys were boys, and kumquats were kumquats. Besides, I think your son is going to explode if he has to sit still any longer.'

They wander through the crowds and over to a bazaar table, while Matty runs around tables, bumping into people as heedlessly as if they were some kind of soft obstacles, then instantly changing direction.

At the bazaar table, they hand over some change, and receive a handful of paper twists, which they take turns opening. The prizes are set out on the table—religious bookmarks, plastic rosebuds, china figurines.

'You've won something,' says the young woman to Matty, helping him to open his paper twist. He presents the scrap of paper at the table, hopping up and down until he receives a small plate with an insignia on it.

'Congratulations,' says the woman. 'It's the Lisboa coat of arms.'

Back at the table, Eduardo takes the plate in his bandaged hands, turning it over.

'Are these crows?' says Geneviève, pointing to two birds perched above a boat on the plate.

'Ravens,' he says.

'Over a corpse ship,' says the young woman to Geneviève. 'Isn't that appealing?'

'Really?' says Geneviève, laughing. 'That's the coat of arms?'

'Guarding the corpse, though,' says Eduardo, as if this explained it. 'Or escorting it. Not scavenging it, anyway.'

'Whose corpse?' says Geneviève, trying to straighten out her face. She lifts Matty onto her lap, where he curls up and starts sucking on the collar of his T-shirt.

'Saint Vincent. Patron saint of Lisbon. He was martyred in Valencia, and then brought by ship to Lisbon. His body was supposed to be miraculously preserved during the trip.'

A preserved saint. She thinks of him folded into a large preserving jar, surrounded by amber liquid. Or rolled out flat and dried, like fruit leather. But surely the saints are supposed to preserve us?

At least he's Catholic, said her mother, when Geneviève presented her with Eduardo. Even if he was one of *les autres*.

Tu es Catholique? This was her mother's first question to the friends she brought home, children from the other end of the skipping ropes. A redundant question, too—they were all Catholic, fully, comprehensively Catholic.

Ah, her mother would say lightly, as if it hadn't mattered, now that she knew.

Look at the Protestants, Geneviève's father would say to them, glancing slyly at her mother. They have it easy. They're out playing while you are doing *catéchisme*.

Rotten wood, her mother snapped anxiously. Better not to even talk about them.

But Eduardo's Catholicism has a dry, sinewy quality that Geneviève has never seen before. It seems to occupy a remote part of his brain, as if he has been able to corner it into one spot. She envies this, his ability to contain something she finds so intimate, so messy. How did he extricate himself from that swampy mix, that feeling of being cherished and punished at the same time?

Geneviève rubs her hands together to wipe off some bread crumbs. Matty is almost asleep on her lap. The night air is cooler now, a reminder that summer would be departing shortly—loitering sometimes, as it often did in Montréal, and then rushing off suddenly, as if it were late for an appointment in another hemisphere. But for the moment, summer is still here, and she is content to sit there with the strings of lights glowing above them, to feel satiated, watching the faces of the people around her, listening to a supple language she doesn't understand.

She hears the word *explosão,* and glances over at Eduardo. He must be talking about the fire yet again—he is speaking quickly, intensely, his eyes fixed on the plate, which he is still turning over and over in his hands, the painted ship glinting, Saint Vincent making no objection. Dreaming in his preserved way about Lisbon, safe in his patronage. Except for the odd earthquake or plague.

Is there a patron saint against fires? There must be. A patron saint of architects? Maybe. She remembers Eduardo talking about the Protestant architect who designed the Basilica at the corner of Notre-Dame and Saint-Sulpice.

He was so taken with his own work that he converted to Catholicism, said Eduardo. So that he could be buried there.

At the time she laughed, and he looked surprised. To her, this seemed the height of absurdity. To him it seemed perfectly plausible.

The band starts up again now, plodding through a piece of music. It sounds lugubrious, and she turns to Eduardo, ready to mock it. Then she sees that he is listening intently, his eyes half-closed.

She looks around again. People have collected on the sidewalks beyond the courtyard, watching the revellers, curious about them. She feels suddenly generous, suddenly expansive.

Come in, she wants to say. Don't be shy. Come in and win a coat of arms. Wouldn't that be handy? Wouldn't that be useful? Don't stand out there. Have something to drink. Have something to eat.

Then she remembers that this isn't her party.

Chapter Six

'Fill this out,' says the receptionist at the sleep clinic, handing Patrick a questionnaire. He notices the other people in the waiting room are writing on clipboards as well. They seem strangely intent, as if putting the right answers down might alone solve their problems.

On a scale of one to five, how lightly do you sleep? On a scale of one to five, how refreshing is your sleep? On a scale of one to five, how much do you dream?

One. One. One.

Are you fatigued during the day? Do you have difficulty staying awake during the day? Do you have difficulty falling asleep? Do you wake up at night gasping for air?

Yes. Yes. Yes. Maybe.

He hands the clipboard back, and looks around at his fellow non-sleepers.

Why can't any of you sleep? Why can't I sleep? Is this an epidemic? Is someone stealing our sleep, hiding it?

Not that he hasn't tried to figure this out himself. Maybe he stumbles into the wrong night when he gets into bed. If he had a nocturnal map, he might be able to find the right night, the one that contains sleep, instead of a room full of stale thoughts, a room in a constant state of disarray.

No, this is backwards—night isn't the producer of chaos. Khaos was the void that produced the Greek goddess of night, Nyx. And Nyx was the mother of Hypnos, the god of sleep.

A god of sleep? How suitable. Drowsy, surrounded by the Oneiroi, the tribe of dreams. Except that this god seems to be untrustworthy, failing his duties. Where is he now? Lazy? Lost? Taking a nap?

Or maybe he ran off with his brother—Thanatos, the god of death.

'Patrick Burgess?' says the receptionist. 'The sleep technician will be with you in a minute.'

He studies a framed picture on the wall, faux parchment with remedies for insomnia in antique script.

Wild lettuce
Silk bags stuffed with hops
Sour zizyphus seeds in straw baskets
Powdered topaz in wine
Natrum muriaticum 30X
Running cold water over the ankles
Raw onion on bread
Rubbing fragrant oil on the scalp and the soles of the feet

Maybe some of these would be worth a try—nothing else seems to be working.

Are you worried about something? his doctor had said.

Worried? Not worried, particularly. More of a sense of unease, of pressures building under the surface, of subtle shifts in tectonic plates. And then there is the divorce, and the shower of hatred that followed. But none of this is new, not something that would suddenly produce sleepless nights.

'Come in, please,' says the technician.

He is holding the door open, and motioning to Patrick. They walk down a hall and the technician steers him into another room, dimly lit.

'You can change into your pajamas here,' he says. 'I'll be back in a minute.'

The room smells faintly of salt and iron, old sleep. When the technician returns, he has Patrick lie down on the bed, then he washes spots on his skin and scalp with a gel. After this, he attaches the electrodes, pasting them onto his skull and taping them to his calves, his cheekbones, the skin at the corner of his eyes, behind his ears. The technician does all this so gently and carefully that Patrick finds it unnerving.

The wires are attached to a box, and the technician leaves for another room where he will monitor his alpha waves, his rapid eye movements, his nerves and muscles, his oxygen levels, his heart rhythms.

Patrick waits, lying in the dark, his head feeling hollow. He isn't waiting for sleep—he knows better than this. He is waiting for nothingness, waiting for sleeplessness. The silence, a blank silence, spreads around him, sinking into his skin.

Worried? Maybe *baffled* might be more accurate, still dumfounded by the onslaught of Beth's emotions. Yes, he understands why she is angry, but not so extraordinarily, persistently angry. Where does this inexhaustible emotion come from? Is it being supplied to her by some kind of underground stream? Blame it all on me, and then be done with it, he wants to say. Either that or share the blame. But talking doesn't help—every word he says seems to enrage her.

Can't we be civilized about this? he said, early on. Can't we do this without it becoming some kind of extreme sport?

Civilized? she had said. Is that a code word for making it easier on you?

Isn't civility a virtue in itself? he said. Objectively.

For the complacent, maybe, she said. Civility works for you because everything's fine in your little shadow world.

He had no intention of asking her what she meant. Neither could he explain to her that some truths can be muddy, can coalesce gradually over a period of time. Especially if the person who has possession of them can rationalize his way out of anything, persuade himself of everything. Another occupational hazard.

But he hadn't even been aware of the feeling building within him—it was as if various parts of his brain were meeting in secret and making decisions without his knowledge. Not until the end did one of these decisions push their way into his consciousness, as a fully formed declaration: *I cannot stay here one more day, one more hour, one more minute.*

But why? said Beth.

I don't know, he said.

Then how can you be so certain? said Beth.

I don't know that, either.

He saw Beth looking at him in a strange mixture of compassion and anger, as if her heart were broken that he was such a moron, and at the same time, she was planning to garrotte him at the earliest opportunity.

This was at the beginning, before the urge to garrotte him won out so completely. He suspected that referring to secret meetings in his brain was not going to change that look. How was it that a man whose currency was words, whose profession was words, who was so fluent he had been accused of glibness, was so utterly unable—to this day—to explain any part of this? And even more concerning—what other edicts might the secret plotters be issuing?

One disaster at a time, he thinks.

With Beth no longer there, though, he feels as if he has become more peripheral, someone hovering around the margins of Eduardo and Geneviève. As a couple, he and Beth were a sufficient counterweight—without Beth, the centre of gravity has somehow shifted to the others.

The others? Or Geneviève? Be frank. What is it about her? She seems to have a more intimate relationship with circumstance, with incident, with everything that falls into her life, everything that brushes up against her.

Why is he only noticing this now? Because he wasn't looking before? Or maybe over the last few years with Beth, he has become lopsided, deformed by being half of something, and now he needs something to fill the space, or he will fall over.

This is the standard post-separation prescription anyway: *find someone else.* As if this involved going to the grocery story, picking out a new mate. Or for the more public-spirited—adopting one from the Humane Society.

Of course, Geneviève is a friend—nothing more. Or really a friend—nothing less. And there are a thousand gradations and types of friendship, all with different shadings and colourings. Especially different shadings of sensuality.

He feels his penis stiffening. Will this show up on the monitor, in his oxygen levels, his alpha waves? He tries to suppress it, without effect.

A friend, nothing less. All the categories for the way people are linked together seem suddenly simplistic. Friend, relative, lover, wife—a handful of words. In reality, so many variations, complexities, overlapping layers, all fluctuating continuously. If man were really a social animal, he thinks, we would have more words for this.

Instead, a few anemic adjectives: old friends, new friends, good friends, fair-weather friends, distant friends.

Illusory, deceptive—none of them real, none of them reliable, according to his father. Something he pronounced like a judgment, in his precise, caustic way, convinced that humanity—in both the general and specific versions—was morally rancid. An austere surgeon to begin with, he was slowly soured by humdrum tragedies—a quiet knife-fight of a divorce, a long-running malpractice suit, a stale career—all of this so tightly held that he eventually died from the effort at 57. Or so his daughter—Patrick's sister—claimed. Practical, self-contained, she put up with her father's sarcasm until she married an arc welder when she was 17, and fled to Prince Rupert, as far away from Montréal as she could get without falling into the Pacific.

None of them real. But Geneviève is real—if anything, too real.

Of course this is out of the question. Absolutely. Categorically.

Look how long he has known them. How long is it? Where did he meet them? Was it that dinner party a few years ago? One of those parties where people drink too much red wine and talk ceaselessly—to each other, over each other, around each other. About this politician, that play, this scandal, that book, mocking, praising, laughing, arguing. It was a party he felt he had been to many times before, but he had approached it with anticipation anyway, looking forward to the heightened pitch of spirits.

The night had been steely cold, with the snowbanks frozen hard, the street glazed over with ice. There was a dusting of new snow on top of the ice, which made it even more slippery. By the time he arrived at the house, people were already there, standing around in small groups, glasses in hand. He joined them and watched as more people arrived in twos and threes, stamping their feet, talking about the ice, unwrapping scarves. They balanced on one foot, a hand on the wall, prying one boot off with the toe of another. They piled their coats on the newel post, or slung them over the banister, and then came further into the room, rubbing their hands, touching their cold faces to other cold faces, glad to see each other. *Allô-toi. Bonsoir. Salut.*

Out of the corner of his eye, he saw Geneviève and Eduardo arriving, her hair mussed from her hat. When they had drinks in their hands, Eduardo crooked his elbow around her neck, and began telling a story.

À table, said the red-faced host distractedly, a serving bowl in each hand.

There was a tacit sorting process as they moved towards the table—people drifting towards certain chairs, hesitating, noticing imperceptibly who was drifting around them, and either sitting down or changing course. Then they were settled, and he found himself next to Geneviève, Eduardo on the other side.

At the table, the smaller conversations of the party merged into larger ones, divided again, merged again in different configurations.

It should be possible to diagram this, Geneviève said to him in French. All the individual voices, their combinations, their patterns, the outline of the whole thing. Like a piece of music.

Are you a composer? said Patrick.

No, just a player, she said.

What do you play?

The viola. A sadly misunderstood instrument. But my daily bread is botany, I specialize in fungi.

You can specialize in fungi? How many are there?

She looked at him in disbelief.

I'm going to report you to the International Association of Mycologists, she said. They'll probably hang you from your ignorant toes. For your information, there are thousands of them.

Mycologists? he said.

No, she said laughing. Fungi, thousands of fungi. But I research patterns in how they reproduce.

Are you being lectured to about fungi? said Eduardo, turning around towards them. You'll have to excuse her—she can't help herself.

It's not lecturing, she protested.

You're right, Eduardo had said. It's more like proselytizing. Recruiting for the Church of Mycology.

The old Eduardo. Where did this new one—the morose, withdrawn version—come from? Was it only the fire, or had he been gradually drifting into this for a while?

You're so *oblivious*, Beth shouted at Patrick once, as if this was the worst crime of all.

But people are entitled to some privacy—this is what he thinks. Constantly scrutinizing the people around him, inspecting them for hints of their inner state—surely this is a form of spying? Or at least eavesdropping on other people's emotions, their thoughts. He certainly wouldn't want someone doing it to him—the idea repels him, makes him uneasy. But these are old thoughts—he has said them so often to himself, they have been worn smooth and polished.

There is no clock in the sleep clinic room, so he lies there, not knowing whether the slippery night minutes have expanded or shrunk. He is overwhelmed with weariness, but sleep is no closer than ever.

They are studying his sleep architecture, the technician said. Eduardo would be incensed at the use of this word. Strange how we acquire things from the people we know–views, gestures, even their irritations.

But this blurring and melding of edges—all the more reason for a degree of enclosure, a degree of shielding—to avoid personal dilution. Not a new thought, either, but still sound, still true. Isn't it? Yes. Maybe. Yes, of course it is.

Are you watching? he says to the technician in the other room, monitoring his brain waves, his breathing, his muscle twitches. Surely the most intimate form of spying. What can you see in all those blips and jagged lines? What can you divine from all those electronic peaks and valleys?

Tell me.

'They're starting,' says Patrick.

He and Eduardo are standing under a night sky, on the flat roof of Eduardo's apartment building. Overhead, the black expanse, diluted with city light, is dotted with white beads. The night has a chilly edge—September has muscled itself in—and the city sounds are softened by the night, the height.

'This better be worth it,' says Eduardo, yawning.

'Over there,' says Patrick. 'Take a look.'

'I don't see anything.'

'Some of them are fainter than others,' says Patrick. 'Keep your eyes open.' Was that really a meteor? He isn't sure himself.

A white streak shoots across the sky.

'There,' says Patrick.

Another streak appears and disappears.

'One at a time? Why is it called a shower?' says Eduardo.

'Sometimes there are several at once, and you can see a flurry of them over a relatively brief period of time. In the grand scheme of astronomy, that's a shower.'

Eduardo sits down on a box-shaped roof vent, his bandaged hands in his pockets, and tilts his head back to look up.

'I hope it gets more interesting than this.'

'If you're bored already, you can take a look at that star cluster over there—that's the Perseus double cluster,' says Patrick. 'And there's Cassiopeia next to it.' The brisk wind ruffles the collar of his shirt, his hair.

'Do they do anything?' says Eduardo.

'No, they're stars, not meteors—they're not required to perform feats of amazement.'

'I'm still waiting for something to happen,' says Eduardo.

'Have you heard from the insurance investigator?' says Patrick to fend off further complaining. 'I don't think I'm up to date on the latest installment.'

'He called me—I'm meeting with him tomorrow.'

'That should move things along,' says Patrick.

'Not fast enough—now I'm getting strange looks from people in my own firm. I can't understand how gullible people are. Or maybe it's not gullibility—how ready they are to believe something absurd. They should know me better than that. I keep wondering if there's something about me that makes this thing more plausible. Some character defect they see that I've never noticed.'

'Well, you do have *arsonist* written all over your face,' says Patrick.

'*What?*' says Eduardo.

'Relax,' says Patrick. 'I'm joking.'

'*Relax?*' says Eduardo, his voice suddenly thick with anger. 'You think I should *relax*?'

'Get a grip,' says Patrick impatiently. 'I'm on your side, remember?'

There is a heavy silence. A night bird hoops in the distance. The air smells like the St. Lawrence—a wild, brackish smell—and car exhaust.

'I know that,' says Eduardo, after a minute. 'I know that. There just seems to be fewer of those people than I thought. At this point, if someone said I was a rapist, I think people would believe it. It's as if saying something alone gives it a kind of life of its own. How could it be so easy to smear someone?'

'Maybe people aren't good at detecting lies,' says Patrick. 'Maybe it's an evolutionary thing. Anyone who ignored *Look, a sabre-toothed tiger!* didn't survive. So we're predisposed to belief. Maybe we haven't adjusted to the falsehood quotient of modern life.'

'But how are they willing to entertain the idea at all? How is it that this seems even remotely possible to them?'

'I don't think it's that personal. There's a constant stream of stories about public figures who turn out to be unscrupulous, politicians on the take, evangelists having affairs. Not to mention insurance fraud. You're being judged as a member of the human race—which turns out to be almost comically untrustworthy.'

A white streak appears and disappears in the dark sky, then another, and another.

'And don't forget, some people lie spectacularly well,' Patrick adds.

'Unlike me—I seem to tell the truth spectacularly badly.'

'There's still the insurance investigation.'

'It better be done damn soon—before the firm goes under,' says Eduardo, and then clears his throat.

'What's that over there?' he says quickly.

'An airplane,' says Patrick. 'But see that brightish star over there? That's Deneb Algedi, part of Capricornus.'

'I think I see it.'

'It's one point of a kind of lumpy triangle shape.'

'Then I don't see it.'

'It's one of the dimmer constellations,' says Patrick. 'Third magnitude at best.'

'I think you make this stuff up.'

'Tempting, given how ignorant you are,' says Patrick. 'But not this time. Capricornus is supposed to be Pan, a minor goat-god who was the son of Zeus. He and Hermes retrieved Zeus's sinews for him after they were stolen. Zeus was grateful, so he took Pan up into the heavens.'

'What happened to Hermes?'

'Became the god of thieves and liars,' says Patrick. 'Much more interesting.'

'Is that the kind of useless thing you learned in that insular private school of yours?'

'Yes, as a matter of fact.'

Two streaks shoot out at the same time, one with a faint bluish-green tinge.

'They look like they're coming from that side of the sky,' says Eduardo.

'That's the radiant point.'

'What does that mean?'

'It's an illusion,' says Patrick. 'Like driving in a snowstorm, when all the snowflakes look like they're coming towards you.'

'It's just a trick of perspective?'

'Don't sound so disappointed. Aren't you're in the illusion business yourself?'

'That's not the same thing at all,' says Eduardo shortly.

'Do you have any idea what a pain in the ass you are?' says Patrick.

Eduardo grunts.

Maybe Geneviève will appreciate this excursion, thinks Patrick.

(I do, Geneviève assures him.)

He looks up at the dark sky, its wavering curtains of stars, its shifting depths. Two white streaks shoot out, one upwards and one to the east.

'What are they, anyway?' says Eduardo.

'You didn't learn that in your virtuously all-embracing public school? They're dust and debris from the tail of a comet. They burn up when they hit the earth's atmosphere.'

Another streak appears, this one tinged with orange. A few seconds later, a faint green one drops suddenly down. Then three streaks flare out, one right after another, in shades of green and blue.

'I thought they would be showier, but I have to admit this is kind of impressive in its own way.'

'And that, ladies and gentlemen, is my point,' says Patrick.

'Are you taking credit for this?' says Eduardo, suspiciously.

'Certainly. You don't think you'd be out here at one in the morning if it wasn't for me?'

'But you can't take credit for the meteors.'

'If a meteor burns up without anyone seeing it, did it really happen? Although,' he says, 'they do drop star jelly in their wake.'

Eduardo snorts. 'You're kidding.'

'Not at all. It's called *pwdre ser*. Welsh for *rot of the stars.* People were always finding it on the grass in the 1600's. Or maybe it was the 1700's. They even wrote about it. *Seek a fallen star, said the hermit, and thou shalt only light on some foul jelly, which, in shooting through the horizon, has assumed for a moment an appearance of splendour.'*

'I don't believe it,' says Eduardo. 'It doesn't really exist, does it?'

'Who knows?' says Patrick.

You're a fraud, said Beth.

Not true, says Patrick. Or no more than anyone else.

Who are you really? she says. Just out of curiosity.

Chapter Seven

The insurance investigator looks nothing like Eduardo had imagined. He is obese, rolls of flesh ringing his body, his clothes—an expensive-looking suit, a light green shirt—containing yards of fabric. He looks far too young as well—his dark blond hair in tight curls, his forehead damp with sweat, despite the autumn weather.

How old is he? Eduardo thinks in dismay. This is the man who will be deciding his fate? Or least his business prospects—which in his case, might as well be his fate.

The investigator carefully lowers his massive body into a leather chair across from Eduardo's desk, an operation accomplished in stages, as he expertly arranges his bulges. He seems to be in possession of his fat, deploying it rather than being surrounded by it or even engulfed by it. This gives him a certain presence, an element of dignity.

'Now,' the man says, wiping his forehead with a handkerchief, and then taking out some forms. 'Let's start from the beginning. What were you doing at the market that day?'

'I went to meet the owner.' Surely he knows this already?

'But the owner wasn't there,' says the man. He is taking notes, his left hand holding the pen awkwardly between his second and third finger, his forearm curled around it.

'I was early, I wanted to look around the market, get a better feel for it.'

The investigator leans forward, his rolls of fat rearranging themselves. Eduardo glances out the window, at a beech tree beside the building, its branches beginning to turn dark gold, waving slowly, heavily in the September wind.

'Did the market owner ever show up?'

'I guess so,' says Eduardo. He hadn't thought of this before. 'Probably. But I wouldn't know, I've never met him, I've only spoken to him over the telephone.'

'Then why would he want you to design his renovations?'

'He said he had seen some of my other work, that he was impressed by it.'

'What other work?'

Eduardo tries to recall what the owner said.

'I can't remember,' he admits.

'So someone calls you up out of the blue and wants to hire you without even meeting you?' says the investigator.

'No,' says Eduardo sharply. 'The idea was that we were going to meet there and discuss it.'

'How did you know that he was really the market owner?'

'I suppose I didn't, but why else would he call me?' Too late, he hears this hanging in the air.

'That's what I'm here to find out,' says the investigator smugly. He leans back in his chair, now displaying his bulk in a way that seems almost calculated, his ivory-coloured tie looking flimsy on his massive, shirted stomach.

'Are you telling me he wasn't really the market owner?'

'Not at all,' says the man, 'I'm not here to tell you anything.'

Eduardo has to restrain himself from throwing his stapler at the man's head. He looks out the window again instead. The gold branches of the beech are tossing more swiftly now, shimmering in the light.

This is asinine, he thinks. Or it would be, if the stakes weren't so high.

But as the questions continue, he realizes that the investigator is cleverer than he seems. His questions circle around each fact, jabbing at it, and then each question in turn is ringed with other questions. And he is thorough—Eduardo feels as if the man is performing an exacting scrutiny of his brain.

'You say you've never met the owner?' says the man.

'No.'

'But you called him a number of times after the fire.'

'Yes, I couldn't get hold of him.'

'How often?'

'I don't know—maybe two or three times over four days.'

'Why were you so anxious to talk to him?' says the man, sitting back again, patting his face with the handkerchief.

'I wanted to talk to him about the commission,' says Eduardo.

'Two or three times over the next four days? What did you talk about?'

'I told you, I couldn't get hold of him.'

'But you did talk to him eventually?' says the man.

'Yes.'

'And did you talk about the commission?'

'No,' says Eduardo. Why did this all sound wrong?

'What did you talk about?'

'The fire,' says Eduardo reluctantly.

'Anything else?'

'The insurance investigation,' says Eduardo, even more unwillingly.

'Ah,' says the investigator.

Cobwebs of smoke begin drifting into the office.

'I understand your firm is in financial difficulty,' says the investigator.

'Not particularly,' says Eduardo, trying to sound casual. 'Or no more than most firms. We can always use more work.'

'I'm going to need a copy of the firm's financial statements and tax returns,' says the investigator. 'Then I may have more questions for you.'

'I doubt you have a right to that,' Eduardo says evenly. 'Is this really necessary?'

'Only if you want the insurance claim to be paid,' says the investigator.

'It's not my claim,' says Eduardo pointedly.

'No, but you benefit from it.'

'We don't even have the commission yet,' says Eduardo, trying to control the exasperation his voice.

The investigator shrugs, the shrug flowing through his bulges. 'Not much chance of getting it if the claim isn't paid out.'

'I'll think about it,' says Eduardo.

'Don't think about it too long.'

Who *is* this man? Eduardo is having trouble reconciling his aggressiveness with his boyish face. Maybe this is his advantage as an investigator—the incongruity keeps people off balance. The smell of smoke is stronger now, and involuntarily, he rubs a bandaged hand across his nose.

'You were seen studying the market,' the man says.

‘Of course I was studying the market—that’s why I was there,’ says Eduardo irritably.

‘I thought you were there to meet the owner.’

‘Both,’ says Eduardo. ‘I was there to do both.’

‘Now,’ says the investigator, ‘this fish vendor—do you have his telephone number?’

‘Why would I have his telephone number?’ says Eduardo. ‘Surely you can reach him through the market.’

‘He isn’t back at work yet. And other witnesses say you came out of the building together.’

‘That was just chance,’ says Eduardo impatiently. ‘I saw him lying on the floor and tried to help him. It could have been anyone.’

‘Anyway,’ he adds, ‘the market owner must have some way to reach him. He would have rented him the stall.’

‘Probably,’ says the investigator, rolling over to one side in his chair to take something out of his briefcase.

Eduardo realizes that the man knows perfectly well how to reach the vendor. The smoke is getting thicker, and now there is another smell—something charred and unpleasant. Burnt fish? He shakes his head abruptly.

‘You’re not seriously thinking this was some kind of conspiracy?’ he says.

‘I’m not ruling anything out,’ says the investigator. ‘Let’s talk about this elusive man, this man you claim threw the cigarette butt. Describe him for me.’

‘I’m not *claiming* anything, I saw him do it,’ says Eduardo. But as he tries to describe the man, he realizes that his impression of him is vague.

‘Do you remember his name?’

‘How would I know his name?’ snaps Eduardo.

Then he takes a breath and tries to steady himself. He needs this man on his side, he needs a favourable report, and he needs it quickly. Before the rumours get any further.

'You were seen talking to him for several minutes.'

'He asked me for money.'

'And you gave him money?'

'Yes.'

'How much?' says the investigator, as if Eduardo had handed over an envelope of cash.

'Just some change.'

'Why did you follow him indoors?' says the investigator.

'I wasn't following him,' says Eduardo tightly. Then he realizes that he actually did decide to go into the building after he saw the man do it—but only because he was hoping it would be cooler inside. He doesn't feel like explaining this to the investigator.

'And you had never seen him before?'

'Of course not,' says Eduardo. Surely the investigator doesn't really think that he hired this man to start a fire? Or that if he had, he would really be so feckless as to admit it?

'But you were only a few feet behind him when the fire started?'

'Yes,' Eduardo says. He has given up trying to explain—this only makes him sound defensive. His eyes are beginning to sting again, and he coughs.

'Has he ever turned up?' he says to the investigator, thinking that he is entitled to ask a few questions himself.

'Why do you want to know?' says the investigator offensively, pen in hand.

'Just curious,' says Eduardo, trying not to react.

'If I were you, I would stick to answering the questions,' the investigator says.

If I were you? All at once, he is infuriated by the man's suspicions, his dismissiveness, as if Eduardo were some shifty operator. He has always prided himself on not being pompous, not being self-aggrandizing, but he suddenly realizes this was only possible because he was usually treated with respect, that it was based on a kind of tacit pact.

Now he feels like exploding at this investigator who seems determined to belittle him, to diminish him with his insinuations. He takes another breath, reminding himself that he needs this man, he needs this report.

'After all,' says the investigator, 'if I find that the fire was deliberately set, you could be a potential suspect.'

The blood rushes to Eduardo's head.

'Get out,' he shouts. 'Get out of my office.'

If I were you, said his father, I would keep an eye out for cork trees.

Spain. How old was he? Four? Five? Hours in an old red Fiat on the way to Ronda, holding a sticky bottle of orange Sumol, his father at the wheel, his mother in the front seat. The monotony of the drive, watching endless stretches of grey-green olive trees rush by, watching the ground change colour from red to brown to yellow and then back again. As his father drove, he pointed out caper bushes, fields of withered sunflowers. The grove of cork trees, several men prying the cork bark off in sections, the tree trunks looking pale and skinned underneath.

His parents' friends—a puffy-faced woman, a man with a guttural voice—were delighted when they arrived in the late evening. So good to see you both, and this is Eduardo? What a big boy. We have arranged everything, we will take you around, show you everything. Have some dinner, rest, and in the morning we will see everything.

They produced pumpkin soup, tripe with beans, custard tarts. Eduardo fell asleep on the floor while they talked.

In the morning, they stood in a row, near the Puente Nuevo, his mother holding his hand tightly. They could hear the Guadalevín churning three hundred feet below, but the gorge was covered in mist, a grey cloud knitted around it.

We will have breakfast, announced the man. We will show you the restaurant first, and then come back to the bridge. The mist will have burned off by then.

They ate potato omelettes at the restaurant, and drank strong coffee with milk. His mother and father admired the shape of the bar, the chairs, the granite tabletops, the poster of Oporto on the wall. (See? We will feed them, then send them on to Portugal.) The man, stocky, mild, listened to them while he flipped through a pile of invoices. His wife fished for compliments, and then waved them away.

When they returned to the bridge, the mist was still there, but it had separated into wisps. The gorge was a deep crack in the earth.

Look at that rock, said his father.

Look at that water, said his mother.

Look at that *bridge*, said Eduardo.

How did they make it?

Ask the architect, the man who built it three hundred years ago. A project that took thirty years.

The adult Eduardo cannot imagine this now. Surely there were moments when the man lost heart? Or interest?

When the bridge was finished, the architect had to examine it, to inspect it. Standing underneath it, looking up, did he feel a leap in his chest? Did he feel a rush of grateful dizziness? Did he stand on the bridge and stroke the roughness of its stones?

He died there. This is the story, that he fell from the bridge into the gorge. One version is that he killed himself, after completing his ultimate life's work. Another is that he slipped and fell.

Of course, both versions are untrue. He died later in Málaga, from illness.

Still, architecture has its hazards.

He looks around at the apprehensive faces in the boardroom. He has called a meeting of the firm—to clear the air, he thinks, but the truth is that he is feeling the need for allies as well. People are still coming in, senior associates, junior associates, interns, some standing against the wall—the colour of the week is a deep burgundy—some sitting down. The shelves along the side of the room are stuffed with books of building materials, and there is a clump of vegetables on the table. One of the interns who lives on a farm has taken to bringing in fall produce—tomatoes, peppers, zucchini in the hope of getting rid of the excess.

I'm at my wits' end, he says.

That didn't take long, says another intern.

Honestly. You can only make so much chutney and ratatouille. Vegetables have brought me to my knees.

He has used toothpicks to construct a tableau—a pepper is balanced on a vertical zucchini, a line of cherry tomatoes is sitting on a prone cob of corn, a fan of snow peas at one end, a carrot trapeze hanging at the other. A note is propped at the bottom with a line from a seventeenth century poem: *My vegetable love should grow / vaster than empires and more slow.* Although the intention was clearly sardonic, there is something almost a little wistful about it.

'This will only take a minute,' says Eduardo.

He tells them about the insurance investigation, he makes light of the rumours, he reassures them that this will blow over soon.

But the atmosphere in the room is unexpectedly chilly.

'The library clients are nervous,' says Jeremy aggressively. 'And I've heard it from other clients, too.'

A ripple of agreement spreads around the room.

I should have done this earlier, Eduardo thinks, surprised by the resentful tone. It was a mistake to leave it this long. Of course, they're afraid they will be tainted by the allegation as well.

'Look,' he says. 'I didn't raise this before because I thought it was silly, that the thing would blow over in a day or two.'

'And now you're not so sure?' says an intern anxiously.

'I'm sure that this will die down eventually, but it's taking longer than I thought.'

'We need something to say,' says one of the senior associates, fiddling with a pencil. 'Something definitive that will put it to rest.'

'If only I had something like that,' says Eduardo.

They look at him with alarm.

'No, no,' he says hastily. 'I mean, the allegation is so off the wall—there's nothing to get hold of, that you can point to. It's like trying to prove something in a vacuum. Or trying to prove a negative. What can you say, except that it's not true?'

'But you're always talking about the financial problems,' says Jeremy boldly.

'You don't believe any of this?' says Eduardo.

'No,' he says, uncertainly.

'For Christ's sake, you know what a stickler I am,' says Eduardo, raising his voice. 'And even if I had somehow lost every ethical bone in my body, do you really think I would be so stupid?'

Several of them look uncomfortable.

Has everyone gone insane? he thinks.

'I'm going home,' he says to his secretary. She almost drops her coffee.

At home, Geneviève and Matty are out. Thank God, he thinks. He feels crowded by people, exhausted by people, as if the world had suddenly become too congested. The autumn air itself seems chilled into something heavier, more complicated. Solitude—or even emptiness—feels like a relief.

He lies down on the couch and closes his eyes. But he is too restless to sleep, his mind running over the interview with the investigator, changing his answers, polishing them up, as if this would make any difference now. If only he could send the man this improved version, telling him to discard the first one. How could he have been such an idiot as to lose his temper like that? As if things weren't precarious enough.

And the smell of smoke again. Could this really be his imagination? Or simply stray cigarette smoke? But this is a different kind of smoke, a tricky, malignant smoke. He is reluctant to say anything, to ask Geneviève: *can you smell that?* Reluctant to see the look on her face, that scramble of concern and wariness.

Pull yourself together, he thinks. The firm has work at the moment, the prospects are not so dire. The children's library, the university field house, the waterfront cultural centre, the music school renovations. But these are all in different stages—some on the verge of completion with only the deficiency inspections to go. And in the works? An outpatient clinic, a gallery, several competitions they are waiting to hear from. That damn mausoleum.

Working on the mausoleum design seems to have brought out another side of himself, a patchy, metaphysical streak. Maybe this is a function of getting older as well. He can think of several philosophical propositions that were once charming little puzzles, and now feel more pertinent, too pertinent—and not necessarily in pleasant ways.

For one thing, this project has yielded too much information about the facts of death. He is surprised that some of these details aren't more revealing—he had assumed that they would be more illuminating about death generally. Or maybe he is resisting the conclusion they suggest—that there is nothing more than the body, and the body disintegrates. And disintegrate it does—now he knows all about it. The cells that die, the micro-organisms that break down the dead cells, the enzymes and chemicals released by this that digest the body further. The skin that will blister and colour as a result of gases, the tissues that will eventually liquefy, the body cavities that will burst open. He knows that some tissues take longer to deteriorate, that the prostate gland—odd, this gland in particular—may last for several months.

These details are at least possible to grasp, even if they are relatively useless. The liminality of death is what he finds horrifying, the idea of a free fall into shapelessness, the dissolving of a being. What could be worse?

Perhaps being inconsequential in life.

He is not a believer in tunnels, white lights, beatific afterlives. Stories, metaphysical slush.

What have you got against stories? says Geneviève.

They're not true.

He thinks of the investigator again, the ease with which the man was able to construct an accusing story. At least the interview was over and done before they had to leave for Portugal. Maybe Portugal will be a respite, maybe it will be warmer. Lisbon, the light bouncing off the white buildings, the river—smooth, broad, seamed with green, grey, yellow, blue. Flowing lazily by, headed for the North Atlantic. Lazily? Not lazily, exactly—calmly indifferent to all the bits and pieces of human activity on its surface. He feels suddenly hungry for this landscape, as if it was something that could be consumed.

He looks around at the apartment, full of saturated colours that seem overwhelming. Whose idea were they? He can't remember. The oxblood was his—that he knows. The turquoise—probably Geneviève's. Of course, that framed poster on the wall was hers as well—she had put it up first when they moved in. An artistic composition—red starfish-like fungus, a honeycombed morel with black edges, ripples of golden jelly-like substance, striped brown scalloped fans, a stalk with a white lacy veil, an orange latticed cage fungus, a cluster of phosphorescent green caps.

They had spent most of the day hoisting furniture, emptying boxes, positioning lamps and bookcases.

What do you think? she said when they took a break, sitting down on the sofa to survey their work. She put her bare feet up on a coffee table, trying to balance her drink on her knee—it teetered for a second before she caught it.

Good, he said.

Has anyone ever mentioned that you're not much of a talker? If we're going to live together, you might want to get a little chattier. Just so our home life isn't one long soliloquy on my part. Here, let's practise—I'll start you off. Tell me something about yourself. One, two, three, go.

So he did. He talked again about the buildings he loved, the church in northern Portugal, the rectory in Alicante.

Their lines, poetic and ascetic at the same time. Their blinding whiteness, their impossibly beautiful precision.

I can't really describe them, he said finally. You have to see them.

I believe you, she said, impressed.

That afternoon—was it the first or second afternoon in this apartment? The streets were cold, the winter sun so bright that being indoors in the middle of the afternoon—drinking—felt vaguely immoral. They talked and drank, and after an hour, she got up, took out a plate and a short knife and picked up a pear. She sat back down again, and began peeling the pear, cutting into the skin, sliding the knife slowly along underneath it. A shaft of sunlight from the window glanced off the knife, but she was absorbed in her peeling, her dark hair hanging down beside her face.

When she had finished, she passed a piece to him. Then she uncurled herself from the sofa, and went over to close the curtains. A crack of light remained, disappearing as her body moved in front of it, and then flashing again as she sat down.

Now, talk about yourself, she said, instead of buildings. Here we are, moving in together, so tell me something I should know.

Instantly he was at a loss for words, scrambling to think of something to say about himself that wouldn't sound vain, or self-deprecating, or banal.

Like what?

Anything, she said. Everything. How about Portugal? Do you miss it?'

Miss it? This hardly described his relationship to a city that surrounded him as a child, that filled his pores, his ears, his lungs, his mouth. He had been a small extension of Lisboa, so immersed in it that he had no sense of otherness, no perspective. If he were required to think about other countries in school, he imagined them merely as versions of his own life, a series of different Lisboas. When his apartment, his city, his school, suddenly disappeared in an airplane ride, this life was switched off abruptly. Miss it? He carried bits of it around inside his head for years, something he took out from time to time and examined, as if it were a game—a miniature whitewashed building, small almond trees, a tiny pan of frying sardines.

Maybe I did miss it, he said, feeling a certain stale sadness for that boy, his former self. Or maybe he missed a time when there was nothing to miss. When he goes back now, Lisboa seemed to him to be harder, sharper, dirtier. Sometimes he sees his childhood city, a familiar doorway, a corner that produces an echo in his mind, but this makes him feel disoriented, as if his memory and the present reality were on different glass slides, one on top of the other, but out of alignment.

He had taken one of her hands in his palm and studied it, the thin bones beneath the skin of her fingers, her uneven fingernails, the calluses from the viola strings. He traced the bones from her wrist to her knuckles, flexing her fingers gently. This is a hand I live with now, he thought. This is a hand that is in some way mine. The senses in his own finger pads seemed heightened, and his body was tightening, tensing, desire rushing through it.

The crack of light from between the curtains lit up some strands of her hair. He picked them up, and rubbed them between his fingers as she turned her head towards his hand. Then he began passing his hand slowly over and around her face, under her eyes, around the corners of her mouth, the nerves in his fingers sending out instant jolts to the rest of his body. Underneath her heavy hair, her neck was slender, with a small tail of dark hairs at her nape.

He felt a tremor of excitement as he put his arms around her, the feel of her skin that day—chalky, electric. Underneath her shirt, he touched the faint silvery streaks on her breasts. He felt her hands floating across his neck, his chest, sliding towards his belt, until the hairs on his skin were standing up. Sensation overwhelmed him, filling his mind until it crowded everything else.

He had pulled her over to him, and their arms and legs became tangled. They were still awkward, then, it was all still new—bumping into each other, his neck stiff, her elbow in the wrong place. Sorry, they whispered to each other. Sorry. They adjusted themselves, shifted around limbs, tried to move more gracefully, only to become clumsier.

Gradually, their lack of co-ordination turned into an irregular fluency of its own, one that involved sudden bursts of sensation, and then pauses for rearrangements. When he moved inside her, he tried to adapt himself to her movements, to become part of her rhythm. But she was trying to fit herself to him, and they ended up with the same unsynchronized motion, unpredictably, agonizingly erotic.

He opened his eyes to look at her face below him. Her own eyes were closed, but the rest of her face was so expressive that she might have had tiny words at the curve of her cheekbones, at the corners of her eyebrows. Then her features tensed, her body stiffened, and a spasm of pleasure rippled across her face. He watched as her expression dissolved, and then gradually reassembled with a different vocabulary.

What is the wits' end? says his mother.

And how do you know where it is? says Geneviève. Just out of curiosity.

Chapter Eight

The cottony night air is the first thing Geneviève notices as she steps out the door of the Lisbon Airport—they have flown backwards from late September into summer weather again. She and Eduardo join the queue for a taxi, the terminal behind them, a long sweep of thrumming light. The line moves quickly, and in a minute a driver is slinging their bags into a car trunk.

The taxi races through the dark, taking the city hills so fast that she feels her stomach dropping. At the top of a small rise, they are moving so quickly that she feels as if they will sail out into the air on the other side. The ride is unexpectedly exhilarating, clearing away the frowziness of their flight.

How is it they don't have more accidents? she thinks, just as the driver sideswipes another car. The driver stops and is out of the taxi immediately, examining a dent in the fender, swearing at the other driver, while the other driver delivers his own impassioned views.

'Should we be doing something?' she says to Eduardo after a minute. 'Helping to uphold the honour of our driver?'

'You're easily enlisted,' he says shortly.

This could be a long trip.

The argument between the drivers becomes louder, and just as she thinks they will start punching each other, it subsides. This happens several times, and she realizes that the level of aggression in this argument is carefully calibrated. In a minute, the taxi driver is back in his seat, ranting about the stupidity of the other driver to Eduardo.

In a few more minutes they are at his cousin's apartment, in the Graça district of Lisbon.

'*Entrem, entrem,*' the cousin says delightedly. An ebullient, gregarious man with a brown, lozenge-shaped face. His wife and sons are visiting her parents in the Algarve for a couple of weeks, and he is glad to have company.

In a moment, he has them sitting down with glasses of *porto* in their hands. Over their protests—they already feel overfed from the flight—he brings out a plate of oily black olives and cheese.

He doesn't speak French, but he has some broken English which he uses recklessly, far beyond its limits. Sometimes he seems to be simply tossing out words, but he does it in such an animated way that it almost works.

And who says that communication has to be so precise? thinks Geneviève. This is just a different type. More ambiguity. More possibility.

What does he do? she had asked Eduardo.

He works in an industrial glass factory. Some kind of supervisor. They make everything from fruit bowls to test tubes.

But his job seems to be the least of him, she thinks now. Not something that defines him in any way, that gives him his identity.

She finds this refreshing. Maybe she has been spending too much time with academics. *Cogito ergo sum.* I think, therefore I am—the vocational version.

The cousin seems to be good at—what? At living? At savouring things? At being? At not being hijacked by the need for achievement—although maybe his wife cares about this less than Eduardo does. And maybe he doesn't have to prove anything.

She is overtaken by a series of yawns.

'Sorry,' she says.

'*Não, não,* notatall,' says the cousin in his language blend. He shows her to a bedroom, and she throws off her clothes and crawls into the too soft bed, leaving the left side for Eduardo. Through the door, she can hear the comfortable rumble of talking, the husky sounds of Portuguese. She hopes this is strengthening for him, or comforting, or whatever it is he needs so that he will be less fragile, less irritable in the morning. Then she goes to sleep so fast that she almost collides with it.

In the morning, he takes her to Castelo de São Jorge, so that she can see the lower city spread out below. The white stucco buildings, the terra cotta roofs stretch down to the water of the Tagus River in the brilliant light.

'What makes it so bright?' she says.

'Closer to the equator?' he says.

'Yes,' she says dubiously. 'But not that much.'

Could it be refracted differently? she thinks. Or maybe it has something to do with the reflection from the water.

She stands there, fascinated by it, trying to drink in as much of it in as she can, as if she could store it in her cells.

'What a place,' she says, inhaling the air.

'What a place,' says Eduardo ruefully.

Lisboa. *Alex Ubbo. Olisipo. Felicitas Julia. Al Aschbuna. Lissabona.* A city with too many lives. Conquered by waves of strangers who insisted on renaming it. *This city is ours*, said the Phoenicians. *Ours*, said the Greeks. *Ours,* said the Romans, the Visigoths, the Muslims, the Spanish.

An unpleasant human habit, she said once. This urge to conquer.

Any conquest in particular? said Patrick innocently.

I don't think you're in a position to gloat, she said.

But Québec was only conquered once or twice, Eduardo had said. Poor Lisboa.

When she has had enough, he takes her to see the Monastery of Hieronymites.

'I wouldn't have thought this was your kind of building,' she says. This ornate, Manueline style is as far as it is possible to get from his own.

'Too elaborate and fussy,' he says. 'But it's extraordinary as a piece of history. And you'll like the refectory.'

He leads her into a long, airy room with large windows. The lines are more graceful here, emphasized by the vaulted ceiling. The bottom half of the walls are tiled in yellow and blue *azuelos*, and the light that falls from the windows is quiet and old. The result is a kind of tranquil emptiness.

'The monks must have come in here to rest,' she says, looking around.

'To eat, actually. But I know what you mean.'

The church is not particularly restful, a riot of stone garlands, thistles, fish, ships, poppies, seashells, oak leaves, seaweed, pearls, filigree, all looping exuberantly around carvings of saints and kings.

'Not an age of restraint,' says Eduardo dryly.

'It's so full of stories, though,' says Geneviève. 'It's like a building that's in a constant state of narration.'

He glances sharply at her.

'He thinks I don't know anything about architecture,' she says to a stone king.

They stop at the tomb of Luís de Camões, on one side of the main hall. A sculpted figure of the poet is lying on top of the crypt, his elegantly waved hair on a stone pillow, his hands pointing upward in uncomfortably suspended prayer. Across from this is the tomb of Vasco de Gama, a prone statue in flowing stone robes, and upstairs is King Manuel himself.

'They buried poets together with explorers and kings?' says Geneviève, laughing.

'Do you have to laugh at everything?' he says, irritated.

'Yes,' she says, simply. Especially now, she almost adds.

Then he takes her around the city, avoiding the usual sights to show her smaller, more hidden things—a mossy, blackened fountain, a square scattered with pigeon feathers and pine cones, an old prison, washing lines on balconies, the purple bougainvillea overflowing window boxes, the wash of morning light on stucco walls. He has her look down at the intricate pavement mosaics, and then up at the blue Moorish tiles on the outside of buildings. Then they take the tram out to the Cemetery of Pleasures.

Really? she says, stifling a gurgle of laughter.

It's the local cemetery, he says. *Pleasures* is the name of the area.

Settled by flocks of hedonists? she murmurs to herself.

After this, there is a public washroom in Terreiro do Paço square.

'Only you would consider this a tourist attraction,' she says kindly.

The next stop is Bairro Alto, where he shows her the Igreja de São Roque, with its simple elegance.

'The outside is *estilo chão,'* he says. 'Plain style. Sixteenth century. It's built on the site of a cemetery for plague victims.'

Another cemetery. But his interest in this building makes more sense—this style, its cleaner lines, is more like his own. Inside, though, they are back into gilded woodwork, grapes, birds, cherubs.

'Not so plain here,' she says.

'No, no, this is baroque. There were three different architects involved. Baltasar Álvares was the best one.'

That evening, they end up in a café his cousin recommended. They order pork and clams, but the café is hot and they are still jet-lagged, so they pick at the food in silence. She tries talking several times, but his answers are monosyllabic, so she gives up. He is drinking steadily, but it only seems to make him more morose. Despite this, she feels gripped by a form of inertia—as if they were stuck here, at this table, on these chairs.

Then a *fadista* stands up to sing, along with her guitar players.

'Let's go,' says Eduardo abruptly, pulling out money to pay. 'If I'd known there was *fado* here, I wouldn't have come.'

'Isn't that heresy?' says Geneviève. 'Aren't you supposed to love the poignancy, the melancholy?'

'No,' he says. 'It's maudlin, not melancholy.'

She goes with him reluctantly, on principle, but she is at least partly relieved—whether maudlin or melancholy, this music does seem dispiriting.

It seems strange to think of him as an inhabitant of this country, as if she had discovered another person inside of him. It must have been stranger still for him to have left it so young, to have been grafted onto another country, a piece of branch cut from one tree and fused to another.

But is this any less complicated than feeling vaguely counterfeit—too English to be entirely French, far too French to be English?

And why is it that this mixture seems inferior to being purely one thing or another? Why does she feel she is only half of something—lacking in some way? Why don't these two halves make a whole? In fact, why doesn't she feel as if she is two wholes—twice as good instead of half as good?

We're all half something, says Patrick.

Go away, says Geneviève.

Is it possible there is a finite amount of being, that only a certain amount of source material can be poured into it? No, that seems nonsensical. Or is this merely a failure of imagination, to conceive of new forms of identity? New forms of hybridity?

Mongrels are hardier, Raymond said once.

But more likely to get autoimmune diseases, said Luc. If we're talking about dogs.

The next day Eduardo rents a car and they drive north towards Oporto, where he shows her a modern tea house in Leça da Palmeira, then a bank in Vila do Conde, with its flowing curves, glass and white marble. Álvaro Siza, he says, with unexpected reverence.

Now these buildings, these really are Eduardo, she thinks. These pure lines, this devotion to shape and place and light. And in fact, he seems to revive around them slightly.

She tries to imagine doing something as mundane as her banking at a building like this. Would she get so accustomed to it that she would take it for granted, wouldn't notice the building any more? Or was part of the extraordinary presence of the building that it would never become invisible, that it would always strike the user in some way—perhaps not as forcefully as the first time, but in other, more subtle ways?

Their next stop is another Siza building, a church in Marco de Canaveses, its curved white walls so spare, so quietly dazzling that it makes her bones ache.

'I'm not sure I can take any more,' she says. Is it possible to be over-satiated with this stark beauty?

'Just a few more,' says Eduardo, absently. 'One in Oporto, and some on the way home tomorrow. But we're back to the sixteenth century, and they're not quite so breathtaking.'

'All right, but if my head actually bursts, it's your fault.'

When they finally arrive in Oporto, they find their hotel, and then walk to Igreja São Lourenço.

'Baltasar Álvares again,' he says, pointing out a detail here, an anomaly there.

'Good for him, but I think I'm getting churched out,' she says. 'There's only so much of this I can take without thinking cynical, anti-ecclesiastical thoughts.'

Go ahead, says Patrick.

Go away, she says.

So in the evening, Eduardo takes her to a concert—Wagner, Barber, Stravinsky—played by the Orquestra Nacional do Porto.

'Tell me about this,' she says, as the orchestra tunes up. The paper program is in Portuguese.

'It's a Koolhaas building,' he says. 'Not bad.'

'The program,' she says, laughing. 'I meant the music program.'

The next day, they head south again, stopping in Coimbra to see several other buildings. But she finds her interest still flagging; there is only so much architecture—church or otherwise—that she can take in at one time. And she is starting to feel more remote herself, as if her own state of mind was involuntarily tuning itself to his—but without any increase in connection between them.

That night, she lies in bed, listening to him sleeping beside her. He is sleeping heavily but restlessly, mumbling something every once in a while. His words are indecipherable, but she isn't surprised. In his sleep he speaks a near language, not English, not Portuguese, not French. Words that work their way up through a number of layers within him, becoming distorted along the way. These words have all the inflections of a language, the rhythms, so close that she originally thought if she listened carefully enough, if she strained her ears, she would understand it.

Is this eavesdropping? Maybe, but if he talks a few inches from her ear, she feels she is entitled to listen—in fact, she doesn't have any choice. But at the moment, she would also welcome any information from his depths.

Tell me what is happening to you. Tell me why you seem to be teetering in slow motion. Tell me whether this trip is helping. Tell me anything.

As it turns out, though, he has taken care of this problem, even in his sleep.

She slips out of bed, and wanders into the bathroom, stooping to get a drink from the tap. Then she sits down on the tile floor, leaning against the bathtub. A nightlight glows on the wall, creating strange shadows out of a bottle of lotion, a can of shaving cream.

She begins running through a familiar route in her head, a fertility landscape, full of dead ends. In one spot, her ovaries, conferring endlessly with her hormones—in another, her recalcitrant uterus.

Ah, that uterus. Unruly, or maybe merely incompetent—a bumbler of an organ. The whole system apparently bewildered, perplexed by this most elemental of functions—although to be fair, requiring a finely tuned set of triggers and reactions. A sequence that finally gives rise to the implausible moment of genesis—the eager egg, the dogged sperm, delivered on the wave of an orgasm, merging together.

Or perhaps not delivered on the wave of an orgasm. The studies are still equivocal, although even the ones downplaying the usefulness of orgasms in bringing about this happy conjunction are still breezily pro-orgasm—*have one anyway*!

Orgasms—she has forgotten about her orgasm taxonomy, she is neglecting it. Maybe she should be thinking of more entries.

Paroxysmi studui. Busy orgasms. These are brisk, irritating, hurried orgasms, all sharp corners. They build swiftly, and explode intensely in a shower of black jet beads.

Paroxysmi argenteus. Silver orgasms. They climb slowly, with delicate, silver-blue claws, until a lily-like rush from the roots, and then a burst of green and blue sparks at the top in a brilliant umbrella of sensation.

Is there one more likely to result in a child?

Three years of this increasingly discouraging quest, and a crooked little thought has begun slipping into her mind: how will she know when to give up?

Crooked? Traitorous.

But how *will* she know when to give up? There is no marker, no obvious point when she should stop, when she should reconcile herself to adoption or childlessness. Without some marker, maybe she will keep on indefinitely, led by an increasingly stringy optimism, some confidence in the eventual justice of events. A confidence that is entirely suspect, entirely unsupported—more than that, a confidence that is refuted at every turn.

Of course, the fact that Eduardo is out of commission at the moment (how long a moment?) is not going to help their chances. His hands are free of bandages now, but they are still scabbed over.

It's only your hands, she wants to say. Who needs hands? These aren't the relevant body parts anyway. Do fungi have hands? No. Do lichens have hands? No. And yet look how prolific they are, absolutely handless.

Clinical? Yes, but this is such a clinical process now. Although if she is taking drugs that have risks of mania and delirium, there must be some element of adventure to it.

Stop trying, her mother says. *Then it will happen.* But this doesn't make any sense to her, or at least any more sense than Eduardo's mother saying *a watched pot never boils*. She understands that her mother—more people than her mother—think that giving up has some physical effect, but there doesn't seem to be any real evidence of this. And the truth is that she doesn't know how to stop trying, what steps this would involve, how she would even go about subtracting *trying* from her consciousness. If she tried to stop trying, she would still be trying to do something.

Stop thinking, then, she says to herself. If you can't stop trying, at least stop thinking. For a few minutes, anyway.

The tile floor feels cool under her legs. It will be October when they get back—a sudden shift from the warmth of Lisbon. She rubs her finger along the grout between the tiles. There is something on the underside of her thigh, and she feels it with her hand. A few grains of reddish sand are sticking to her skin.

Eduardo leaves this everywhere, on the floor, in the bathtub, in bed. At first, she assumed the sand was from a construction site, but he seems to acquire it even at his office—and now even here. She sometimes wonders whether he produces it himself, the way skin cells flake off, a by-product of a man who is a builder, a generator of things. A kind of manufacturing dust from his thought processes. He works so laboriously that the idea of him leaving some residue behind isn't entirely absurd.

Of course, he has some excuse for being so laborious, she thinks. What a job—mixing physics and aesthetics with the most persistent of human yearnings, the most practical of human needs. And his buildings—buildings that are so uncluttered, so exquisite, that they sit on a thin edge of something. Temporality? Actuality? Mortality?

No wonder he talks in his sleep.

Then again, maybe he is making up for how laconic he can be during the day, something that makes her impatient at the best of times. Her thoughts slip out of her so easily that she wonders why his take so much effort.

You must have a high mineral content, she said to him once.

What does that mean? he said. She shrugged.

She thinks of Patrick's engaging volubility, imagining his miles of words stretching out across the landscape, piling up against the side of fences and houses and then overflowing them, spreading across fields of stubble, flooding into thickets. But his words seem light compared to Eduardo's—maybe they make up in quantity what they lack in poundage. She can't help wondering if Patrick acquires people with his words, winds his words around them in warm, subtle loops. Is he a collector of people, a consumer of people? Is this what drove Beth to fury? But she isn't Beth.

She yanks her thoughts away from Patrick, back to Eduardo.

What if Eduardo is getting worse instead of better? There are certainly no signs of improvement. He needs to build something, buildings are his cure, his remedy. They provide him with a sense of durability, she thinks, something to counteract his feelings of transience. Maybe they are announcements of his existence, his continuance. Or maybe they weigh him down instead, maybe they congregate in his head, looking down on him with disapproving stares.

She hopes he has too much momentum, too much stubbornness to let the fire get him, to be buried by the arson whispers. Sometimes she wishes she had more of that momentum herself—she often feels as if she is only riding along in her life, watching events fall into it from time to time. He certainly thinks that she should be more ambitious—and maybe he is right. Although the idea of using her research as a vehicle, instead of an end in itself, bothers her in some undefinable way. As if the fungi were not enough in themselves, their strangely beautiful forms, their earthy habits. The first land plants, four hundred million years old. Isn't that astonishing enough? Isn't that amazing enough? Imagine the luck—the sheer, astounding luck—of spending so many hours a day wandering through these miniature forests.

But sometimes she wonders if her lack of ambition is more a matter of cold feet than anything else. Of course, she doesn't really believe in Raymond's bad gene—it wouldn't be a gene anyway, it annoys her the way people throw these terms around. But maybe there is something else that runs in the family—some insidious, subterranean belief, maybe a superstition. What would it be? That pride really does goeth before a fall, that ambition inevitably comes with its own trip wires?

No, no, says Luc, it's more like being attracted to failure, having a weakness for it. *Seduced* by failure, he says, pleased with this idea.

Just an excuse, says Marie-Thérèse, rolling her eyes.

The oldest and the youngest. Well, naturally they don't get along—they smoke different brands.

What do they argue about?

Everything and frequently. They started young—Geneviève remembers one of their fiercest. A cottage, a place near Mascouche rented by an uncle, although it was really more of a cabin. Was it on an island? They had to get to it by boat, anyway, a rowboat with rusted oarlocks. When they arrived, hot children tumbling out of the car, the boat was on the other side of the river. They stood on an old dock with one corner submerged in water, while her mother and Marie-Thérèse unpacked bags from the car. Then the bags were lined up on the dock beside them, bags with roasted chickens, thick bread, potato salad, coconut suntan oil.

Look, said her uncle, showing them a metal basket, big enough to fit two people. It ran along a cable, stretched over the narrowest part of the river, a makeshift gondola. One of her brothers climbed into it—a younger Raymond?—and he cranked himself across, swaying and jerking in mid-air. On the other side, he climbed out and went down the shore to get the boat, tied to the tall stump of a pine. Then he rowed back to the dock, to pick the others up. It took three trips with the boat to ferry across all the children, all the bags.

The first afternoon, Raymond tried to teach them how to dive. She and Paulette lined up docilely in the sun, their toes curled around the end of the dock. They bent their knees and bodies, and then held their hands together over their heads like the prow of a boat. On his count, they fell stiffly forward into the dark green water.

Her father sat on the dock in a folding chair with sagging plastic straps. Beside him, Marie-Thérèse lay on her back on a towel, her arms and legs slick with oil, holding a paperback. Her father applauded each dive, then told them what they had done wrong.

After several rounds of this, they lost interest, and spent the rest of the afternoon splashing and wriggling with the others, their dark hair plastered down on their bobbing heads.

Like otters, said her father.

The next day it rained, though, and they sat in the cabin and played *le pendu*—hangman—on scraps of paper. When they were tired of this, they played cards, *jouer aux 8*, stopping frequently to bicker about the rules. Then suddenly, Marie-Thérèse and Luc were in a shouting argument about whether Luc was cheating.

Her father ignored them pointedly, the others were enjoyably aghast, but a little frightened. After a minute of shouting, her mother took a pan of rice pudding out of the old oven and slapped it down on the table between them, making the cards jump. Then she picked up a knife, slowly, deliberately, and cut a hole in the crust of the meringue on top of the pudding. She took a battered pot off a burner, gave it a shake, and then poured a burnt sugar sauce into the hole, so that it spread out under the meringue. They watched in silence, mesmerized as she dished out bowls. Only when they had finished eating did they start talking again, but by then they were glassy-eyed with sugar.

Marie-Thérèse and Luc still argue, but now they are more likely to snipe at each other indirectly, to other family members. Isn't this more grown-up, isn't this more mature?

Geneviève's legs are getting cold, sitting there on the tile floor. She can hear the cousin snoring delicately in the other bedroom. She stands up, and wanders back to their bedroom, sliding under the blankets, feeling the warmth from Eduardo's body lying next to her. Then she crooks an arm into a half-moon on her forehead, as if the weight could press her thoughts into better shapes.

We're all half-something, says Patrick again, putting his head down under the surface of her sleep.

Half-something, she thinks, in a dream-drunk way.

Half-wise. Half-broken. Half-rare.

Chapter Nine

'What about here?' says Patrick in French, several weeks later, swinging Chloe off his shoulders.

'Fine,' says Eduardo absently.

'Gorgeous,' says Geneviève, dropping her bag, and untangling herself from Matty.

They are standing on a granite outcropping in a wilderness park near St. Jerome, looking around with the satisfaction of arrival. On one side, a narrow river is surging over boulders and rock ledges. On the other, the terrain falls sharply, spreading out below them, a vast expanse of trees, their leaves glowing—swaths of pale reds, patches of rusts and apricots, fringes of clarets and yellows. The colours seem almost cocky, tempting a fate of some kind. Which of course, they are, Patrick thinks—the last brilliant, daring flush before they die. Colours with nothing to lose.

Patrick watches Geneviève as she straightens up and takes a deep breath of the clear air.

'What is it about the trees turning that seems exhilarating?' she says. 'I mean, it happens every year—we should be used to it by now, blasé. Is it the sense of expectancy? The idea of being on the brink of something?'

'I take it this is a rhetorical question,' says Patrick.

'I guess so—unless you know the answer, in which case it isn't.'

She was the one who had proposed this fall picnic as a way to try to lift Eduardo's spirits, to pull him a few centimetres out of his gloom. But don't tell him that, she said—I told him it was to get Matty outside and run him around.

There is no doubt that Matty needs running around.

One of Marie-Thérèse's pins worked its way out, so she's still laid up, said Geneviève. We have him for a few more weeks.

Something flies by Patrick, and he looks over to see the boy hitting the ground with a small branch. Chloe trots over to watch him in thumb-sucking admiration.

Geneviève shakes out a tablecloth, which billows out when the wind catches it, and refuses to settle down, despite her flapping.

'Here, grab an end,' she says, laughing as it wraps around her, and Patrick catches a corner and they wrestle it down to the ground. Matty bounds into the middle of it and collapses cross-legged into a sitting position.

'Perfect,' says Geneviève. 'Just what we needed—a tablecloth weight.'

Chloe tries to follow, and trips over the folds, then crawls to sit near him.

Geneviève begins pulling out containers and parcels of food.

'Listen up,' says Patrick to the two children. 'You're not just a couple of pretty faces—you're an essential part of our unique food distribution system. Now Geneviève will pass the food to Matty, who will pass it to Chloe, who will pass it to me. Nod your heads if you understand.'

They nod their heads.

'More of a pipeline than a distribution system,' says Geneviève. 'And don't think we haven't noticed where it ends up.' She starts passing things to Matty—cold ham, mustard, pears, fruit tarts—who promptly begins opening the parcels, ignoring Patrick's instructions.

'This is scandalous,' says Patrick. 'Our food network has broken down, all as a result of one bad apple.'

Matty picks up a corkscrew and begins waving it in Chloe's face.

'*Non, non*,' says Geneviève, disarming him, and ripping off pieces of baguette for both children to hold them for the moment.

'I say we abandon our distribution model and take a look at this river,' Patrick says.

He picks up Chloe under one arm like a bundle, and pulls Matty up by the hand, leading him down to the bank. At the edge of the rushing water, Matty starts poking at reeds with his stick. Chloe watches from under Patrick's arm.

The water is heavy, swirling around massive slabs of stone, flowing down a steep incline in the riverbed, a waterfall that has been stretched out horizontally, white spume in the air. The sun reflects off the wet surfaces of the rocks, black, grey, veined in places.

'Any fish there?' says Patrick.

Matty leans over to study the water, poking more energetically. Then his foot slips, and Patrick grabs him by the back of his shirt. He lifts the boy up and deposits him further up the bank, where he begins shaking his wet foot.

'You're a walking personal injury claim,' says Patrick. 'You could make some civil litigator very happy.'

He puts Chloe down and herds both of them back to the picnic.

'A soaker, a time-honoured tradition,' says Geneviève, patting Matty on the back. She takes off his wet shoe and sock, wrings out the sock, and puts them both on a rock to dry in the sun. Then they settle down around the tablecloth, Eduardo's knees creaking as he sits down on the ground.

'You need to start running again,' says Patrick. They haven't run on Sundays since the market fire.

'For once you might be right,' says Eduardo.

They fill their plates, although Matty ignores his, lying on his stomach, inspecting the ground. As they eat, he uses his piece of baguette to makes lines of dead pine needles. Then he sits up, one leg in an L-shape in front of him, the other behind him, and begins pressing the palms of his hands into the grass.

The whine of a chainsaw starts up faintly in the distance.

'Did you see the news clip about the man who took a chainsaw to a house?' says Geneviève lazily.

'Where?' Patrick says, abruptly turning around.

'Pointe-Claire,' she says, surprised.

'Oh, Pointe-Claire,' he says, relieved. 'Probably a disgruntled architect.'

'Actually, there *was* an architect who used chainsaws,' says Eduardo. 'He cut into existing buildings, removing parts of ceilings and floors, turning them into enormous sculptures.'

'I wonder what happens when you cut out walls with a chainsaw,' says Geneviève. 'Does it leave phantom walls? Like phantom limbs after an amputation?'

'Not unless walls have nerve endings,' says Patrick. 'Which seems a little unlikely, despite what Eduardo might think.'

Eduardo grunts.

After a few minutes, Geneviève puts her plate down, and leans back against a tree, her legs stretched out in front of her. She tilts her head up to the sun and closes her eyes.

'God, this is luxurious. It doesn't seem October.'

She holds out her glass, eyes still closed.

'Here,' says Patrick to Matty, giving him a second bottle of wine. Matty lugs the bottle over to her and refills her glass, splashing it on her hand, her legs.

'*Merci*,' she says, wiping her hand on her jacket. She gives Matty a fruit tart and relieves him of the bottle, and he puts the tart into his pocket.

A chipmunk darts near them, then hovers.

'Shouldn't they be hibernating or something?' says Patrick.

'Too early,' says Geneviève. 'And they only partially hibernate.'

Matty offers the chipmunk a piece of mashed fruit tart. The chipmunk darts closer suddenly, and then skitters off. The boy stumbles backward, sitting down abruptly.

'And that concludes the natural science portion of the trip,' says Patrick, as Geneviève picks Matty up and tries to clean the tart off his hands.

Then they sit in the sun, the boy curled in Geneviève's lap, sucking on part of his shirt collar. The sound of the water is unexpectedly lulling. Chloe is already asleep, stretched out next to Patrick on the tablecloth. Eduardo alone is looking at the river as if it had deliberately offended him.

A few minutes later, Patrick shakes himself.

'Somebody talk,' he says. 'I've had too much to drink, I need to hear something riveting, so I don't fall asleep. One of the cardinal rules for insomniacs—don't take naps.'

'I heard something interesting about some cellists the other day,' says Geneviève dreamily.

'Is it riveting?'

'I think so.'

'Well, then, by all means, let's hear it,' says Patrick.

'There are these three cellists,' she says, and then stops.

'Good beginning,' says Patrick, encouragingly. 'Unless that's the whole thing, in which case it's a little short,' he adds.

'No, no, there are three cellists who play their cellos in unlikely places. In a two week period, they played on the roofs of all the Anglican cathedrals in England. They've also played on the four highest mountains in Scotland, England, Ireland and Wales.'

'I'm not sure I would call that riveting,' says Patrick.

'You're awake, aren't you?' says Geneviève.

'Barely.'

'I'm going for a walk,' she says, standing up and dusting herself. 'Anyone want to come? There must be some interesting fungi lurking in there.' She gestures towards the thicket behind them.

'Me, me, me,' says Matty, hopping up and down on the spot.

'I know you, you, you,' she says laughing, taking his hand and shaking it.

'Sure,' says Patrick. 'If only to see what could possibly constitute an interesting fungi.'

'Go ahead,' says Eduardo. 'I'll stay here.'

'Can you keep an eye on Chloe then?'

Eduardo looks around at her—she is sleeping on the blanket.

'Fine,' he says vaguely.

Geneviève and Patrick begin picking their way through the trees, stopping to help Matty over bracken, around birch saplings with yellow leaves. Here and there are fallen trees, some caught at skewed angles by the branches of other trees, as if they had crashed down into their unwilling arms. Some have died upright, grey skeleton pines still holding boughs with clumps of brown needles. Bushes with scarlet leaves are scattered through the undergrowth.

After a few minutes, Geneviève stops and crouches down. Matty squats down with her.

'Find something?' says Patrick.

'Spoon-leaved moss, I think,' she says, pointing to a small patch of green tentacles. 'Not a fungus, but still interesting. It's a hard moss to identify though—especially in this light.'

He crouches next to her, suddenly aware of the smooth skin on her face a few inches away, a small scar near her ear, the smell of her dark hair—some spice? He breathes in, trying to ignore the urge to hold her face in his hands.

Stop that, he says to himself.

But would he feel this way if it weren't mutual?

He stands up.

'Well, it's not very impressive looking,' he says. 'But I guess that's the fate of mosses generally. How does this rate on the interesting scale?'

'Well, I think it's endangered.'

'I'm not surprised,' he says.

'You don't sound too concerned. That's not the approved environmental attitude. You're supposed to gasp with alarm at the very thought.'

'What's one moss more or less?' he says. 'There's far too much vegetation here as it is. Can't you see we're surrounded?' He gestures with his arm.

'I think we'll live,' she says, extricating Matty from a clump of bushes, brushing some leaves out of his hair.

'That's only because we can outsmart them—we have the superior brain power.'

'I hope so,' says Geneviève. 'Since they don't have any.'

Patrick bends over to Matty.

'Let's try a disguise,' he says. 'Quick, act like a plant.'

Matty looks puzzled.

'We may have to go for ignominious surrender instead,' says Patrick.

They wander further into the forest, climbing over humps and logs, stopping from time to time for Geneviève to point out something.

Eventually she finds an orange half-circle with a yellow rim on a tree, like a dinner plate lodged in the trunk.

'One of the bracket fungi,' she says. '*Pycnoporus.*'

Matty starts kicking at it and she pulls him away.

'Well, even if this forest isn't dangerous,' says Patrick, looking around, 'you have to admit that it's very messy. Not to mention full of dirt. We need the seven maids with the seven mops in here, pronto.'

'It's elegantly organized at the cell level,' she says. 'You're looking at the forest instead of the trees. Or the leaves. Besides, tidiness isn't everything.'

'You think so?' he says.

'I hope so,' she says.

He feels suspended in this moment, his skin tingling with tension, the rich, dead smell of fallen leaves surrounding them. He reaches for her in a half-gesture, almost involuntarily.

What am I doing? And why am I doing it so badly?

'Let's go back,' she says uncertainly. Her face is flushed. 'Aren't we supposed to be cheering up Eduardo?'

Back at the picnic site, they find that Eduardo has been cheering himself up—the second bottle of wine is almost gone. Chloe is still asleep, drooling gently.

'Find anything interesting?' he says.

'Only fungi,' says Patrick, taking the bottle from Eduardo and pouring the rest of the wine into his glass. 'Which is, of course, fascinating,' he adds quickly.

'Too late,' says Geneviève. 'You've already demonstrated your callous attitude towards fungi. Anyway, you're forgetting about the spoon-leaved moss.'

'Right, the spoon-leaved moss. The spoon-leaved moss was really the climax,' says Patrick.

'Sounds like I didn't miss anything,' says Eduardo.

Patrick is back in his office, willing his clients to disappear. They are starting to seem like a collected mass of legal desires, a tide of needs that he can't possibly satisfy. He wishes he could say to them all, or at least more of them:

I have the perfect solution to your problem. It will only take a minute, and it won't cost you a thing. Now get out.

He has lost interest in them, at least for the moment. His mind is now populated—overpopulated—with Genevièves. Geneviève intent on threading a string through a tuning peg, Geneviève presenting a diagram on the wall of the greenhouse as if she were a magician that had just made it appear, Geneviève holding out a glass for wine, kissing him hello, goodbye, touching his arm. *You're a lifesaver.*

But surely this is a harmless fantasy?

No, even he knows better than that.

He has stationed a sentry to patrol the perimeter around his brain, to keep all of these Genevièves out. Shoot to kill, he says to the sentry. But the sentry can't handle their guerrilla tactics, their nimble strikes, their ability to appear and disappear at will. This isn't an overpopulation—this is an infestation.

He is angry. After the divorce, all the turmoil that it took to get here, he has promised himself some period of idle, uncomplicated sex, a stretch of carnal indulgence. Wasn't that his reward? Yes. Wasn't he owed that? Yes. Why would he get fixated on someone—anyone—but particularly someone who was so completely, utterly, absolutely unavailable? Assuming that this is the case. No, no—the whole thing is unthinkable. Unthinkable even at the best of times, but inconceivable now, with Eduardo in this—this what? This condition? Eduardo is like a man who has been knocked off balance, but is unable to fall—suspended in a state of disequilibrium.

Find someone else.

Look who I found, he imagines saying to Eduardo.

In the meantime, Eduardo has taken up his suggestion to begin running again. The thought makes him wince. But can he really say no now? Put him off? For how long?

And in his expanded list of terms for friendship—*man, the all too social animal*—where would this fit? A treacherous friend? A dishonourable friend? A betraying friend?

You're an idiot, he says to himself.

I *am* an idiot, he says.

This is deranged, he says to himself.

It *is* deranged.

So far, no dispute.

Then how did you manage to get into this position?

Well, I'm an idiot.

On any conventional scale, this would be a capital offence, a rupture of the most basic understanding of trust. At the same time—so commonplace. What moral sense is there in something so widely rejected in reality? But this is an argument and he—of all people—knows the slippery habits of arguments, the insidious ways they coax and persuade. Then how should he think his way through this? By instinct? By touch? What will not be undermined by self-interest and the agility of his own brain?

Of course, he had clients who had fallen foolishly, impulsively, destructively, disastrously for other people. He had been secretly contemptuous, dismissive, even while he reassured them. *Why are you so feckless? Why are you so foolhardy? What could possibly be so extraordinary that you would incinerate the rest of your life like this?*

A judicious man, he had thought of himself. An intelligent, reasonable man. A man whose appetite for risk was low—a virtuoso of a bystander, an expert non-combatant. And law school was no place to learn otherwise. The cases he studied seemed to add up to a long string of cautionary tales—surgeons who amputated the wrong leg, husbands who were caught *in flagrante delicto*, children who played on railway tracks, drivers who took their eyes off the road for a few seconds, and ended up in fiery collisions. His textbooks were full of people floating in a sea of consequences, of inexorable causes and effects.

It's enough to induce a kind of risk-averse paralysis, said one of his classmates.

But his feelings about Geneviève are resistant to any attempt at reasoning, any attempt at self-protection. He is not in love—even this poor, degraded, hollowed-out word, this jack of all trades makes him shudder. No, this is something different—more along the lines of a delirious sense of attachment, his attention is glued to her. Some species of infatuation? Not the seesawing between euphoria and misery of a teenager, but another variety—odder, cooler in temperature, but no less powerful for all that.

Infatuations—those ridiculed things, dismissed as temporary failures of judgment, delusions that will inevitably be regretted. *He'll return to his senses.* What if an infatuation itself actually represents a person returning to his senses? A breaking with the kind of unnaturally prudent, well-modulated existence that people force upon themselves. What if an infatuation represents a person at his most alive? What if it is a method of germination, a beginning?

Or maybe the secret conspirators in his brain have simply mounted another coup.

Are you interested or not? says Eduardo, several weeks later. It was your idea.

I'm interested, says Patrick grimly.

If they are going running, he has to drop Chloe off at Beth's house first. He is tired of these weekend exchanges, the tension, the formality. They make him feel as if he were an under-credentialed attaché with a dispatch for a hostile country.

He finds leaving Chloe difficult as well. He has been ambushed by her. When he sits with her on the sofa, when she curls in his lap, lying against his chest, he is frankly amazed by the tenderness he feels, by the fact that he can sit for so long with this warm bundle sleeping in his arms. He has never thought of himself as particularly fatherly—in fact, he couldn't even describe the elements of fatherliness except in the most basic terms—and he would never have predicted this self-effacement, this strange emotional adhesion, this willingness to flatten himself around her. It seems at odds with almost everything else he knows—a piece outside the puzzle, an additional clue. Something that might even suggest doubt about other things. But she also takes considerable energy and an elephantine patience, so this half-time arrangement offers some relief—despite the fact that Beth makes these handovers as difficult as possible.

She is waiting for them on the porch as he drives up. Time for his weekly brush with hatred.

In the back, Chloe is crooning something to herself around the thumb in her mouth. He reaches over to retrieve her overnight bag. Poor Chloe, they have managed to turn her into a nomad. Although maybe this makes for an interesting childhood.

Beth is stony-faced, animosity flowing from her in waves. She has a tendency to earaches—the things they know about each other—and he can tell by the way she holds her head that she has one now. She says nothing to him, taking the bag from him silently, turning her back on him.

He wants to yell at her departing back: *all right, all right, I'm guilty. Guilty, guilty, guilty. Is that guilty enough?*

Eduardo is already waiting when he arrives at the trail, doing a few perfunctory stretches. They run in different places—Bois-de-Liesse, Île Sainte-Hélène, along the Lachine Canal—but today they are at Cap-Saint-Jacques. The October air is cold and fresh—the wind has blown it clear—and the sky is a reckless blue, thin scraps of clouds thrown across it. The tree colours are darker now—bronze, copper, dark saffron—although they are thinning out, some already bare, patches of grey and brown. They start off easily, their feet thudding on the path. He is shorter than Eduardo, which means they run in a breaking rhythm; first in stride with each other, gradually falling out of step, and then slowly moving into it again.

The path extends along the water, edged with weeds, an old paper cup near the bank, a mass of water lilies trailing yellow stems—their pads faded orange, dull red, edges tattered.

They usually make a point of talking while they run, to pace themselves so that they can talk comfortably. This is the right way to do it, the proper way.

So we can yell *help, I'm having a heart attack*, says Patrick.

But he is tired of hearing Eduardo talk about the explosion, the investigation. Yes, it was bad. Yes, the arson thing is disturbing. But all Eduardo is doing is circling around and around the same territory, picking through it to look for non-existent explanations, stuck on the stubborn injustice of what happened.

Your fifteen minutes of sympathy are up, thinks Patrick.

'I had a look at your contract,' he says. 'It's not bad at all. I'll mark up a copy and send it to you.'

'Thanks,' says Eduardo.

'If only all my clients were like you. Although maybe a little less moody. What's happening with the mausoleum contract?'

'I've come up with some changes to the design, although probably not enough. And we're working on some of the practical issues.'

'Such as?' says Patrick. He is starting to sweat, even in the cool air, and he wipes his forehead with his sleeve. Poplars are edging the path here, the sun catching their lemon-coloured leaves.

'How to seal the crypts properly, how to control the odours. It's not like a grave site. The building has to hold over a thousand human bodies in it, in various stages of decomposition. Then there's the insect issue.'

'I'm not sure I want to know about this.'

Eduardo ignores him.

'The formwork for the cement crypts leaves small channels in the surfaces. They're a conduit for insects, so we have to solve that. And we need to tilt the crypts slightly, so that the fluids run off.'

'I'm absolutely certain I don't want to know about this.'

A thousand human bodies.

The sun is getting higher, and he notices with satisfaction that Eduardo's shirt is stained with sweat as well. He swerves to avoid some sumacs, their branches a burst of crimson, reaching long fingers out over the path.

'You know I run about a mile and an eighth for every mile you run,' he says.

'You're carrying less weight,' says Eduardo, speeding up.

Patrick speeds up as well. They run like this for a few minutes, and he starts to breathe more heavily. Now he is irritated with Eduardo for inflicting this pace on them, and he accelerates more himself.

If we're going to run, let's run.

Eduardo speeds up more in response, and suddenly, they are in a flat-out race.

Bushes flash by and he can hear himself breathing hard, the thud of his feet filling his ears. For a moment, he has the sensation that they are running on the spot, and that the path, the landscape is moving instead.

Another mile, and sweat is trickling into his eyes, one of his calves is aching, the precursor to a cramp. They hit a boggy patch and slow down around it, then they are racing again, submerged in this frenzied contest. The colour-stained landscape is blurring around him—there is only Eduardo and him, their ragged breathing, their grim determination. A branch whips his face, but they pound on, locked into a strange union of effort.

Another mile—two?—and his heart is slamming into his chest wall, his feet have turned into cinderblocks. He is almost praying that something will happen to end this—a sprained ankle, a fall. But then he stumbles himself and a small burst of adrenaline forces him on.

Out of the corner of his eye, he sees Eduardo falter, holding his side.

The finish line, he thinks, but then Eduardo has pushed on ahead.

Now his legs are weak, his head is starting to feel dizzy.

Stop, for God sake, he says to Eduardo in his mind. Are you trying to kill us?

You stop, says Eduardo.

He is nauseated by the time Eduardo finally does stop, bending over, his hands on his knees, gasping for breath.

He forces himself to take a few more strides, raising his arms in the air in shaky victory, then he lurches over to a log beside the water to sit down. He uses his shirt to wipe his face.

'I must be drinking too much,' says Eduardo, between gasps. Then he straightens up, and heaves himself over to the log as well.

They sit there, exhausted, in strained silence, their heart rates bumping down slowly, sweat drying on their skins. Close to the bank, under the shade of a yellowing willow, a dispirited duck is paddling around. Farther out, the sun has caught the angles of the ripples, so that it looks as if someone has scattered crushed glass across the cold water.

After a while, the duck dives down, and then surfaces farther out. There is a soft thunk near the bank, and something disappears, leaving circles of ripples. A kingfisher gives a rattling call, and swoops across the water.

Patrick yawns violently.

'Still not sleeping?' says Eduardo.

'I'm waiting for the results from the sleep clinic. I'm even taking some herbal remedy now.'

'What is it?'

'I can't remember the name. My father is probably rolling over in his grave.'

The duck disappears and surfaces again, this time holding something in its bill.

'Where is your father's grave, anyway?' says Eduardo.

'You certainly know how to hold on to an idea,' says Patrick.

Chapter Ten

'First the good news,' says the market owner. 'They found the man who threw the cigarette. They've decided that he doesn't have the brains to pull off some scheme, that it was just stupidity on his part. Although the police are still thinking about charging him with something.'

Res ipsa loquitur, thinks Eduardo. Apparently the thing really did speak for itself.

You're welcome, says Patrick.

Eduardo feels almost sorry for the man—one small, everyday act, something he must have done a thousand times. But then he shifts the telephone and flexes his hand, the skin shiny and pink where the scabs have fallen off.

'Anyway, they've decided it was nothing to do with me. Or you. So the insurance company is paying out on the claim. I should have the money shortly.'

Eduardo is overwhelmed with a sense of relief as tangible as an atmospheric shift, as if an oxygen current had begun flowing around him, restoring a normal colouring and order to everything, blowing away the greyness that has settled into his body, his bones.

'Do they make some kind of announcement?' he says. 'Do they issue a report?'

'Well, no,' says the market owner. 'They've just notified me they'll be releasing the funds.'

The oxygen current stops flowing abruptly.

'Then how will people know that the arson thing was untrue?' says Eduardo.

'They're just rumours. And the insurance being paid says something. You can spread the word yourself, if it bothers you so much.'

It doesn't bother you? What kind of an idiot are you?

Aloud, he says: 'Let's set up a meeting as soon as possible then. My office is useful because we have samples of building materials here, but really, any place that's convenient for you is fine.'

'Except the market,' he adds, with a strained laugh.

'Well, that's the other piece of news,' says the market owner. 'I've decided to run a design competition.'

'What?' says Eduardo.

'I'm running a competition,' says the owner more loudly.

Eduardo can hear the words, but they seem like small objects the man is dropping into his ear. Then the meaning clicks in, and he feels a cold rage spreading up the back of his head.

After all that? he wants to shout. After what I've been through?

'Are you still there?' says the owner.

Eduardo takes a deep breath.

'After all that?' he says, in a throttled voice.

'That's one of the reasons why a competition is important,' says the owner stiffly. 'It will make it clear that everything is above board, transparent. And we'll do it quickly, no pre-qualification, an open competition, conceptual designs. And the deadline will be in a month, so it won't delay things too much.'

A month? Eduardo wants to shout. A month? That's insane. A project like this needs at least three months.

Instead he takes another breath. Clearly there is no point in arguing with the man—the decision has been made.

'Who's on the jury?' he says.

'It's just someone from the city planning department, me, and an architect from Boston.'

'What kind of jury is that?' exclaims Eduardo, before he can stop himself.

'Look, I understand your frustration,' says the market owner, more apologetically now. 'But the city's kicking in some infrastructure money now, so that changes things. And besides, there are other architects interested in the project.'

'Who?' says Eduardo grimly.

'Steven Hegge, for one,' says the owner.

Eduardo is filled with a dirty, convoluted anger—an anger without any of the comfort of self-righteousness. How can he complain about the owner holding a competition? He himself is a promoter of competitions, a believer in public processes. A person who argues against the genteel corruption of cronyism every chance he gets.

Not to mention the fact that he relied on the uncertainty of getting the commission to support his innocence, at least with the insurance investigator.

But competitions need juries, and this is barely a jury. The Boston architect is a journeyman, capable in his own way, but pedestrian. What does the owner think he is doing? Is this some half-baked compromise, some deal for the city money? The city should know better. Maybe their bargaining power is limited, their financial contribution not large enough to insist on more than this. And the owner probably wants to keep control over the design.

But he has been assuming all along that if he could come up with a design that appealed to the owner, the commission would be his. He has more than earned it, given everything he has been through.

Not at all, says Baltasar Álvares. The only thing that matters is the brilliance, the integrity of the design itself. If your design is outstanding, you have nothing to fear from a competition. Except, of course, a better design.

Hopelessly naïve, says Eduardo. Look at the competitions where lesser designs have won, where aesthetic biases or backscratching have prevailed over elegance, boldness or originality. And think about this: will the jury be reluctant to choose almost any design of mine, out of concern that it will look as if the outcome was a foregone conclusion?

The idea that Hegge has had a hand in bringing this about is almost incomprehensible. Clearly he had underrated him as a threat. That would explain skipping the pre-qualification stage as well—Hegge wouldn't make it through. And maybe the short deadline—he has probably been working on his design for weeks now.

He is even beginning to wonder if Hegge was the original source of the arson rumour, instead of merely a conduit, although this seems to suggest a degree of planning, of cunning that is beyond his capabilities. It seems more likely that the man had simply taken advantage of the idea that the fire was suspicious to plant doubts, to manipulate the market owner. But then again, he has obviously underestimated him. He can imagine Hegge musing out loud about the fire, planting the suspicion, and then stoking the rumours that followed. That little weasel, that shallow, underhanded purveyor of mediocrity. Why isn't the triteness of his designs obvious? Why isn't the weakness of his character apparent? What could he have possibly said to the market owner to worm his way into his confidence?

You might want to save all that energy for your work, says Álvares dryly.

At least there is one good thing—the cemetery director called shortly afterwards to set up another meeting for the presentation of the new mausoleum design.

Not that there is new mausoleum design yet.

I think we have a solution to some of your concerns about investing the building with a more sacred quality, he had said to the director, untruthfully.

They weren't my concerns, I can assure you, said the director. I've always pushed for your firm, I've always been impressed by your artistic vision.

Despite the fact that the man was flattering him, despite the fact that he had probably called after he had heard about the insurance decision, he found this almost moving. Is this what he's been reduced to?

Self-pity is never an attractive characteristic, says Álvares. Go and design something.

So he does.

Or at least he tries to.

One, two, three, go, says Geneviève.

He is sketching hastily. He has a vague idea that his problem with the market is one of freezing up, that the stakes involved have induced a kind of lockjaw of the mind.

Forget all that, he says to himself. Just throw yourself at it, and see what happens. So he is rummaging through his inventory of design concepts, throwing down any ideas that come along, and then tossing the sketch aside and starting a new one. He has no particular confidence in this system; in fact, it is completely counter-intuitive. His usual approach is a process of incubation, where he pushes an idea through various iterations of refinement and nuance, looking for traces of brightness, volatility, surprise, tension—anything alive and different—to develop further.

Of course, he has had problems before, but never like this. Occasionally at the very beginning of the process, he wonders whether he will be able to summon up yet another design. Even at the best of times, he has to persuade himself into it, to lure himself with the memory of other buildings, to remind himself that all he has to do is catch the tail of an idea, an idea that will grow. To build himself up to a certain level of engineered confidence—a necessary fraud of nerve and belief.

But then the idea takes hold of him, and his thoughts begin firing in a rush, coming so fast that they start overrunning each other. When this happens, he has to stand up, to move around—the intensity of it is draining and heady at the same time.

Nothing compares with the feeling of looking at the finished building, though. He marvels at each one as freely as if someone else had designed it. More than once, he has been tempted to encircle a building with his arms, lay his face against the side, something completely at odds with his usual reserve.

This project, though—he can't afford even the normal incubation process. This is an emergency—a month is far too short for even a conceptual design competition. He has pulled a senior associate and three juniors off other projects, and they are starting the site and program analyses, while he tries to come up with the design. The structural engineers are juggling things around to accommodate us, says Sandrine, although they gave me an earful.

Several hours later, his table, the chairs and even part of the floor are covered with sketches. But they have evolved, any inspired parts that emerged have surfaced in subsequent sketches, and the whole thing has gradually coalesced and become deeper and more intelligent.

And now he has something, something he hopes is striking, compelling. He has used a stylized version of the stone patterns of Lisbon streets to create islands and paths in the plaza—the islands for the stalls, and the paths flowing around them, both echoing the ways people move, and serving as ushers as well. The fruit and vegetable stalls themselves have been designed to fit the plan, each slightly different, but variations on a theme. They also function as plantings, the greens and colours of the produce standing in for the shrubs or small trees that he might normally use. Along one side of the plaza is a long, rectangular water feature—a formal version of a stream—something that reflects the relationship between markets and riverbanks, while the other sides are open and inviting, merging the plaza with its surroundings. The paths sweep up and into the building, where the stone designs continue on the ground floor, behind a glass facade. Inside, the stalls have been laid out in a way that will also shape the queues of people waiting to be served. The second floor has been spiralled around a central atrium, so that the light from the glass ceiling falls into the ground floor in shapes that form part of the arrangement of the stalls and paths.

He calls in the hastily assembled design team for a brainstorming session.

'Here,' says Eduardo, handing them the latest sketch. 'What do you think? First impressions.'

They study it carefully for a minute.

'Well?' he says impatiently.

'Very handsome,' says the senior associate. 'Very good-looking—a European feel. Lots of movement. Great use of light, water.'

Eduardo feels deflated.

'But?' he says.

The associate hesitates.

'But?' says Eduardo again.

'But not particularly adventurous,' says Sandrine reluctantly. 'Your stuff is usually a little more daring.'

'Adventurous? Daring? How can I do that when I have all of you to support?' Eduardo snaps. 'Would you rather I dared this firm into bankruptcy?'

They sit there, mute.

Learn this now, thinks Eduardo. Architecture isn't your own little aesthetic playground.

Now is not the time to play it safe, says Álvares.

Daring doesn't win competitions, says Eduardo.

Difficult to say, says Álvares. And at least you will have a design with its own integrity, not one that panders to the design climate.

A design that will never be built, says Eduardo. I already have too many of those. I have to be strategic, particularly now.

Álvares sniffs.

It's not pandering, says Eduardo. These things are always a negotiation, a balancing act. It's not that simple. And this is a business.

Álvares says nothing.

And this is the way that it's going to be, says Eduardo grimly. But he knows there isn't much point in hiring talented people, and then ignoring them.

'Sorry,' he says awkwardly to the design team, who are exchanging glances with each other. 'Let's assume this is the basic idea, and start batting it around. Maybe there is some way we can push the edges of it, at least to some extent. And let's hear what you've got so far on the site and program analyses.'

'Are you nervous?' says the young man, adjusting his glasses, their round frames dominating his small face.

'No,' says Eduardo. 'It's just a lecture.'

He is sitting in a small anteroom, a few days later. The young man hovering over him works for the architecture museum. He seems almost disappointed, as if he had been looking forward to reassuring Eduardo.

'I'm hoping for a good turnout,' he says, looking at his watch, and shaking his head in a way that suggests this is unlikely.

Eduardo says nothing. He is thinking about the lecture, refining the first few sentences in his head.

'I'll come back in five minutes, shall I?' says the young man.

'Fine,' says Eduardo absently.

The lecture is on Gordon Matta-Clark, his chainsaw architect—or artist. Well, this is one of the controversies. A man who cut out ceilings and walls to reveal different lines, to change the play of light. This might have been an ingenious one-idea wonder, he thinks, but the man was, in fact, talented—and genuinely obsessed with metamorphosis. Although his own interest in him has more to do with Matta-Clark's fascination with orphan spaces, ambiguity.

'Ready?' says the young man, back with a bottle of water.

'Ready enough,' says Eduardo.

The lecture theatre has sixty or seventy people in it. The young man introduces him, summarizing his buildings, his awards. Normally Eduardo doesn't listen to introductions, using the time to assess the crowd, the space, how loudly he will have to talk, the possibilities for questions. But today he finds the introduction soothing, an analgesic for his bruised self-image.

Of course, his self-image and his reputation should be separate things. He knows this. But his public self is a commodity, a professional business asset. He is required to pay attention to it, to protect it. And here he is, doing just that—although this lecture had been scheduled several months ago. He had almost cancelled, the market project swallowing up everything in its path, but thought better of it, deciding it was necessary for rebuilding the firm's credibility.

The introduction is finished, and he stands up, moving to the podium. He adjusts the microphone upwards, and clears his throat. There is always a moment before he starts to talk at one of these events, a few seconds where he feels outside of time, suspended in a transparency of heightened consciousness.

Suddenly, he *is* nervous—not merely nervous, but paralyzed. The smoke has started to drift in again, but this time it is full of other smoke—metallic smoke, grilled meat, cigarette ash, car exhaust, oily smoke, all mingled together. The crowd begins waving their blistering red arms, the plastic bags still melted into their wrists, shouting, their eyebrowless faces contorted with anger. He shakes his head and blinks hard, and they subside into a puzzled audience, uninjured, waiting for him to begin. He plunges hurriedly into the lecture.

He talks for half an hour, describing the way Matta-Clark reversed backgrounds and the foregrounds, the man's sense of excitement about vacuums. His embrace of decline, of the inevitability of deterioration, and his experiments with cracking a building in half. Then he talks about his forays into other areas, his attempts to fry photographs in cooking oil.

He is so caught up in what he is saying that he is surprised to find himself at the end of the lecture, surfacing to a light round of applause. But he leaves hastily, afraid that the crowd will turn into arm-waving burn victims again.

Why is the fire still tormenting him like this? Why can't he shake it off, why does he still have this feeling of disorientation, these unnerving flashbacks? He is embarrassed by this weakness—as catastrophes go, the fire was relatively minor. No-one died, the injuries—his own included—were generally superficial. Of course it was something of a shock, but it should have been a manageable shock—a quick notice of mortality, a salutary reminder to feel grateful for small things, and then a gradual fading of the impact, the details.

But this is not what has happened. Was it the randomness of the fire? If something as ordinary as a market can explode without warning, maybe everything else becomes volatile, suspect.

Or could he be disintegrating in some unknown way? He thinks about his mother, still slowly sinking. Are these things hereditary? She has developed an obsessive interest in cleaning—not someone who cleans over and over, but someone who cleans widely.

She began by sweeping off the sidewalk in front of their house, sweeping until the pavement was bare, clumping on her bowed legs. Then she started sweeping out the small crevices in the sidewalk. Later, she began sweeping out the gutters in front of the sidewalk, and then the street itself.

At the same time, the area of the sidewalk she was sweeping slowly expanded on each side. At first, she was hesitant, taking quick broom strokes, as if she were snatching dirt off the sidewalk that belonged to her neighbours. Then she became bolder, until now she is sweeping the whole block, and part of the next one. She picks up litter, slowly, stiffly, and puts the tops on people's garbage cans. In the spring, she hacks ineffectually at the ice near the sewer grate to make a channel for the snow melt.

He knows there is something absurd about this, but there is something gratingly painful as well—a person who was once capable, whose face has become so blurred, whose brain is calcifying so swiftly.

Her neighbours have responded in different ways. Some are tolerant, especially the Portuguese neighbours, even appreciative. Some are alarmed, as if she had a disease that might be contagious. Various drugs have been tried without success.

At least she's getting some exercise, says Eduardo. And the street is certainly clean. The city should be paying her a stipend.

His father is embarrassed by her, but unable to stop her. What can I do? he says passing a hand over his forehead. I can't tie her down in the house. She is a good woman. She can't help it.

So instead, he spends his time at the garage or the Bar Estoril, playing *sueca*, drinking Sagres beer or *bicas* out of tiny cups with the other men—the ones who look like him, their lined faces, faded shirts, old sports jackets. Talking, jeering good-naturedly as the cards go down, smacking the table with the palms of their hands.

Eduardo can almost see a faint outline of himself at this table, a future shadow, waiting for the right cards to come up. But these are Geneviève-like thoughts—he doesn't have time for them.

Midnight—everyone else is gone. He is examining a coloured rendering of the market building. The table in the boardroom is strewn with cross-sections, floor plans, renderings from different perspectives.

Handsome, yes. Striking, yes. Dazzling, no.

Pragmatism, he reminds himself.

Lack of nerve, sniffs Álvares.

Necessary trade-offs, says Eduardo.

How can you be sure? says Álvares.

I can't, says Eduardo. How do we ever know whether we're compromising too much or too little? There's no gauge, no meter that tells us where that elusive edge is—that this particular point is the furthest we can go without falling off? If there is a fixed edge at all.

Then at least accept that ambiguity.

Shut up, says Eduardo.

He begins picking up the diagrams, one after another, looking for some element that he can develop further. But there is nothing—the design is sealed, complete in itself. There are no undone parts, no entry points—it seems to have an almost muscular cohesion.

Suddenly he is overwhelmed with rage—at the design, at the market owner, at himself, at the whole unbearable thing. He roars with fury, sweeping the papers off the table, knocking over mugs, sample books.

Then he grabs a pad, and begins to sketch feverishly.

An hour later, he calls Sandrine.

'Get the design team in here,' he says. 'I need you.'

'What's up?' she says, sleepily. 'It's 1:00 in the morning.'

'I have a new design.'

'Now?' she exclaims, fully awake. 'That's impossible, there's no way we could pull something together in time.'

'Just get them in here.'

In another hour, they are assembled, yawning, their hair mussed, wearing various combinations of sweatshirts, jeans.

He pins up a sketch on the wall, and they cluster around it. Jeremy's face is screwed up, as if he is in pain.

A gangly intern, who is still wearing a pajama shirt, gives a low whistle.

'What happened to strategy, negotiation, pragmatism?' says the senior associate, still staring at the sketch.

'The hell with it,' says Eduardo, wearily.

A few seconds of silence, and then they begin gabbling, laughing excitedly, almost giddy, the intern in the pajama shirt is whooping.

He can't help laughing himself, looking at some of their gleeful, sleepy faces.

'All right, all right,' he says, as they crowd around him. 'Let's get going.'

If they were working hard before, they are working fiercely now. He has pulled everyone else off other projects, juggled other deadlines, farmed out some work so that the whole office is concentrated on the market design. As the days count down, as October begins to run out on them, pulling November behind it, they work through the design, testing it against the site constraints, the spatial and functional requirements. Their faces become shadowed with fatigue, their skin pale and spotty, but they also seem to become sharper and more creative, as if exhaustion had worn away an outer layer, and their thoughts and ideas were flowing unimpeded by the normal barriers.

They're marinated in adrenaline, Eduardo thinks.

'The engineers are having fits,' says a senior associate. 'They say they can't guarantee the stability in so little time, and there's no way they can do the mechanical and electrical.'

'Tell him we'll take whatever he can give us at this point,' says Eduardo. 'It's a concept competition, and a concept is what they're going to get.'

As the deadline approaches, they are possessed by a grim frenzy. And then there is no more time.

The presentation boards have been set up on a table, and he is doing one last check before they are packaged up for delivery. He stands back for a minute, closes his eyes to rid his brain of images, and opens them again, trying to catch a glimpse of the building, as if it were the first time seeing it.

The plaza has become more stylized, cleaner, but the main transformation is the building—now a white form curved in front to embrace the plaza, sweeping upward on the left side. All the lines are exquisitely simple, almost austere. The upward sweep seems to open up the building to the sky, to create a sense of airiness and possibility. There are no windows in the front, but inside, the light pours in from a glass roof. The atrium is still there, but the spiral of the second floor has been canted to one side, echoing the outer curve in an unpredictable way.

'It gives me chills,' says Sandrine earnestly.

'Let's hope you're not the only one.'

He looks at it again.

'Well, it's a gamble,' he says. 'At least there's that. Although I guess it's always a gamble of some kind. Anyway, pack it up and get it off. And send everyone home to get some sleep. We can try to rescue the rest of this practice later.'

Quem muito abarca pouco abraça, says his mother. He who grasps at too much loses everything.

What happened to *fortune favours the bold?* says Eduardo.

'I have to go early to get set up,' says Geneviève, when he arrives home.

She says this pointedly—he has forgotten about her concert, and she knows it. But this is the very last thing he wants to do tonight—he is exhausted.

'I'll meet you there then,' he says shortly. 'You take the car.'

Maybe he can lie down for a few minutes, close his eyes. He sets the alarm clock for an hour, but he is too keyed up—he feels as if he is staring at the inside of his eyelids. His neck and shoulders are aching from spending so much time hunched over his desk, and he rolls over restlessly, his brain filled with darting thoughts. Finally, he manages to doze for a few minutes, only to be roused—groggy and irritable—by the alarm. Time to go.

He takes a taxi to the house—what had Geneviève said about it? Something he is supposed to notice about it, to remember. The large living room is almost full, people already milling around. Rows of folding chairs have been set up, theatre-style. He looks for Geneviève, but doesn't see her. The instruments are in place, though, so she must be somewhere else in the house with the rest of the quartet.

What do you do before a concert, anyway? he asked her once. Deep breathing? Scales?

We argue, she said. We fight over the vibrato here, the intensity there, the timing some other place. We should have it all figured out—we do have it all figured out. But the nervousness makes us testy, and we start refighting things we've already settled. It's astonishing that we can perform at all, that it all comes together.

Normand, the white-haired owner, claps his hands for attention.

'Asseyez-vous s'il vous plaît, mesdames et messieurs.'

Eduardo finds a chair and sits down with the rest of the crowd. There is a moment of expectation, and then Geneviève and the other musicians walk in and sit down.

The first piece is the Shostakovich quartet. He listens for a few minutes, but he finds it too desolate. He begins studying the room instead, to put some distance between himself and the music, which seems to be pulling at his tired nerve endings.

The wall behind the musicians has a Japanese tapestry of some kind on it, with a thin patch on one side—an antique? The background is pearl silk, with a pale green river running through it, blurred with mist. Two white herons are standing in the river, under clusters of pine needles on branches. Along the side of the tapestry, there are sprays of plum blossoms. He finds the quiet colours, the delicacy of this more moving than the music. Maybe this is because the music is so dark and wild.

The quartet moves on to Haydn, now, though, to the Allegro to the Emperor quartet. He relaxes a little, listening to the notes chasing each other in circles, surprised to find that he is enjoying himself.

A brief intermission, and then they are on to a fierce Grieg quartet. He isn't familiar with this, and he decides that he doesn't want to be. It seems to him that the lively passages are deceptive, always followed by passages of despair.

He tries to concentrate on the music anyway, letting the waves of sound wash over him. He is close enough to the musicians that he can feel the sound through his skin, through his hands. And the notes are so textured, so sensual, that he wonders what it must be like to be playing them, to be in the middle of all that sound.

Finally, they are on to Brahms. (We have to make people happy at the end, says Geneviève.) The first movement is sternly joyous, and he feels a tiny flicker of gladness. It dies again almost immediately, but he is startled at how long it has been since he felt something like that—anything like that.

When the concert is over, they are ushered into a dining room—more milling around a wine bar and a table of green apples, nectarines and cheese. Then Geneviève appears beside him.

'I'm trying to herd you over to meet Normand, the owner of this place,' she says.

'Herd away,' he says.

When they find Normand, he inclines his head towards them, smiling genially. He is talking about an event in Japan a few years ago, where a thousand cellists were assembled to play together.

'It's hard to know whether it would be magnificent or ghastly,' he says.

'Or both,' says Geneviève, introducing Eduardo.

'Geneviève tells me you're an architect,' he says.

Eduardo admits this, and Normand begins talking about design. He seems knowledgeable, and they are quickly into a discussion about buildings they know.

'That coloured glass,' says Normand, shaking his head.

'It's an interesting element,' says Eduardo. 'Look what it does to the light inside.'

His instinct is to rise to the defense of the architect in these circumstances, even when he is not necessarily enthusiastic about the building himself.

'Nevertheless,' says Normand.

'You've had some nice work done here,' says Eduardo, looking around.

'I'm delighted you think so,' he says. 'But I should really go and check on the wine, if you will excuse me for a moment.'

'Of course,' says Geneviève. 'Don't let us keep you.'

'There's another architect here somewhere—a friend of my sister's. I haven't met him myself yet, otherwise I could tell you what he looks like.'

'Don't worry,' says Eduardo. 'I'm quite capable of talking to non-architects as well.'

As Normand moves away, though, Eduardo glances around the room curiously. Who is the other architect?

Suddenly, there is a slack-faced man with sweaty hair standing almost at his elbow.

'I understand we're interested in the same project,' says the man, with an attempt at nonchalance.

Eduardo looks at him, astonished. Wisps of smoke begin drifting around the man, curling around his neck, his shoulders.

'It's a challenging concept, the market,' Hegge says, as if Eduardo had replied. 'We should get together some time, have a drink.'

Is it actually possible that this man thinks they are now colleagues of some kind?

The smoke is beginning to wind around the man's body.

'No,' says Eduardo, his voice strangled, his lungs constricting.

Hegge is transparently offended.

'I don't think you're really in a position to be so arrogant any more,' he says.

Eduardo explodes.

'You little weasel,' he yells, grabbing Hegge's shoulders and shaking him back and forth. 'You little *worm*.'

Hegge's face is disintegrating with alarm.

'Stop it,' he yells, trying to push Eduardo away. 'What the hell are you doing? Stop it.'

'Stop it,' says Geneviève loudly in his ear.

Eduardo drops his hands to his sides, and looks at them as if they were strange attachments he had never seen before. The people around them are silent, delightfully shocked.

'Are you crazy?' says Hegge angrily.

Eduardo shakes his head, trying to clear away the smoke.

Hegge begins stalking away, then turns back.

'If you really want to know who's spreading rumours about you, take a look at your firm.'

'What?' says Eduardo dully.

'Ask Jeremy Boyer,' says Hegge triumphantly.

Pick one, his father used to say, holding his fists stretched out in front of him, facing down. One hand held the *centavo*, one was empty.

He imagines his own fists stretched out. In his left hand—a jumble of flickering images, glimpses of faces, half-built buildings, streets. In his right hand—their unstable variations. Flukes. The possibility of casualties.

Pick one? Pick two.

He wishes he had more hands.

Chapter Eleven

Mont-Royal, a lump of igneous rock, forced through the earth's crust by the cooling and hardening of magma. A geological accident, presiding over a city. Although there are sister mountains as well—Saint-Bruno, Saint-Hilaire, Saint-Grégoire—so perhaps it isn't such an accident after all. But Mont-Royal is a mountain of cemeteries—or at least four of them.

And mausoleums, said Eduardo. I need some inspiration.

I'll go with you, said Geneviève, desperate for Eduardo to be doing something normal—normal for Eduardo, anyway, and desperate for something to do with Matty. Although this may not be an approved children's outing—the three of them, and a date among the dead.

So here they are, a few days later, gloved and jacketed against the chill.

'Cemetery,' says Eduardo in English. 'From the old French *cimetière*, from the Greek *koimeterion*. It means sleeping ground.'

He is studying a small, age-grimed mausoleum with moss growing along one stone wall, one glove off to hold his sketchbook.

'Sleeping ground. That sounds peaceful,' says Geneviève in French. Matty is energetically digging a hole with a stick. 'Careful, you might dig up some bones.'

'Bones, bones, bones,' chants Matty, without pausing in his digging for a second.

A small bird lands on a headstone nearby. *Gloria in excelsis Deo. Marie-Clothilde Pelletier, 1879–1943.* The early November sky is pewter-coloured, and the yellow of a stand of trees—the last to drop their leaves—stand out against it.

'Why do people want to be buried in mausoleums, anyway?' Geneviève says. 'The individual ones I understand—little houses. But the big group ones—they remind me of a chest of drawers.'

She stands up to read another inscription. *À la douce mémoire de Sylvain Gagnon, décédé 23 mai 1951.* Beside it there is a clump of tall grasses, plumes shaped like question marks.

'Family tradition,' says Eduardo absently. 'Cultural tradition. Portugal is full of them. Italy, too.'

'There's something odd about stacking bodies in the air like that.'

She realizes this sounds parochial, but it only makes her feel more obstinate. *Ici repose Etienne Lavallée, époux bien-aimé de Josephine Lavallée.*

'It doesn't use up so much land, though,' he says.

'It uses up space, visual space.'

She moves over to a patch of small white markers, lined up in rows, and stoops to read the inscriptions. The graves of an order of nuns—the headstones are almost nun-like as well. Someone has put a pot of russet chrysanthemums on one.

'Yes, but not as much,' says Eduardo.

'Burying bodies is more natural. They decompose, they're absorbed into the earth.' She breaks off a dead flower stalk.

'They decompose in a mausoleum as well,' he says, flipping the page and starting a new sketch.

'Well, then why not put them into the ground?'

Eduardo turns his head to look at her.

'Is this really necessary?' he says.

She tosses the stalk in the direction of a nun's grave. *Soeur Thérèse-Albert, décédée 9 Janvier 1918. Dei Gratia.*

'I want to be cremated,' she says.

She sees a flicker of contempt cross his face.

A few minutes later, he stands up.

'I'm going over to look at a mausoleum on the other side of the road.'

'I'll stay here with Matty.'

He seems relieved.

She can't really blame him. Why is it every conversation they have these days seems to go sour? She had resolved to be at her best today, or for them to be at their best. But all she had really done was needle him. Although he hadn't done anything to help.

Is this what they are like now? Are they drifting from symbiosis to antibiosis without even noticing? Or maybe the problem is her.

Not likely, says Patrick.

How would you know? she says.

She tries to remember the side effects of the fertility drug currently poisoning her system. Or maybe infertility itself is slowly poisoning her, eating away at her usual levels of optimism or confidence. They should have a blood test for that, she thinks, the same way they measure hormone levels.

Be good, said the nuns in school. Although sometimes they said: *soyez pur*. Be pure.

She stretches out on the grass, underneath one of the yellow-leaved trees. The sun is coming out, first hesitantly, and now more decisively. The branches over her are covered in dark bark, but the leaves are full of light. From the ground, she is looking up into an enormous yellow umbrella.

As she lies there, she feels the deep yellowness enveloping her, she feels herself rising up the middle of the tree, slowly spinning through the leaves, light-headed.

The cemetery is silent except for the sound of Matty digging, but the silence is rich and crowded. All those sleepers, she thinks. All those Pelletiers, Gagnons, Lavallées. Hardy, dogged. *Requiescat in pace.* They deserve their rest, they deserve as much peace as they can find. Or some of them do. Maybe not all of them.

She thinks of her father—still very much alive. When she was a child, when the beer had made him extravagantly tender, he would beckon to her, and lift her gently into his lap, where he would allow her to lie against his warm, round stomach. She could smell his shirt, his hair tonic, the lemon pastilles he had in his mouth. Then he would make her laugh, enveloping her in beery affection.

Sometimes his stomach rumbled. Listen, he would say in his clumsy French, it's talking again. Tell me what it's saying. She would put her ear obligingly against his stomach. *Je sais pas*, she would report. Ah, he would say. Another mystery.

Then he would start mocking the nuns, watching her mother slyly out of the corner of his eye. Name the four gospels, he would say, in a stern, high-pitched voice. *Matthieu, Marc, Luc et.... Hubert,* he would answer himself. *Non, non,* Geneviève said, giggling, and he would pretend to be surprised. Are you sure? Then he would put her down carefully, haul himself out of his chair, and pull a candy cigarette out of a package for her.

Watch this, he would say, holding up the cigarette, showing her how she could smoke it by producing a puff of powdered sugar from the end.

Later in the evening, drunker, he would become spiteful. Then he would snap at her, send her scurrying into another room with a hard knot of hurt in her chest.

Usually when he drank, though, he would drink with her uncles. He would listen as they tried to chart the outlines of this corrosive thing, this Englishness that pressed up against them at every turn. As they tried to take stock of the damage, to add up their losses, to make them more concrete, measurable. As they tried to inoculate themselves with cynicism, or to negotiate a deal with this thing, to bargain for a piece of their own back, only this piece or that piece. But whatever they did, however they tried to protect themselves, to prepare themselves, they kept on getting ambushed over and over.

What did her father think about when they talked so bitterly about the *les Anglais* in front of him—with him? Was there a part of him, despite his odd conversion, that protested? Or did he ever think: *if this is all true, why am I not a luckier man?*

A long sigh of wind sends ripples through the leaves of the tree, changing the light configurations overhead. The rustling slowly builds up, and then subsides again.

An ant is crawling on her arm, and she brushes it off and sits up. She is getting cold. Matty is patting the earth he has dug up onto the top of a gravestone.

Stéphane Thibeault, 1924–1968. Fac pro viribus. Act with all your strength. Matty would agree with that, the way he throws himself at things, whirling through his childhood, driven by curiosity. An ongoing physics experiment, a child of Newton, a careless investigator of the three laws of motion. *Pardon, Monsieur Thibeault.* You don't mind a small boy walking all over your grave, do you?

She moves over to sit with her back against another headstone, watching him. She might as well be comfortable—Eduardo will be gone for a while, drifting from mausoleum to mausoleum.

What an indiscriminate lover he is, she thinks irritably, if lover is the right word. In his own way, he becomes enamoured of things—the lines of a building facade, a theory, a scrap of new information. Although she has to admit he is selective in one sense—only an extraordinary facade, a remarkable theory, a seductive fact.

He becomes engrossed in these things, a kind of communion. He uses them as a sort of source material—they all go into an idea of precarious beauty that he carries around with him, something that is constantly changing. This is what sustains him, she thinks, this is what allows him into the upper reaches of circumstance. But then he is fickle, too—his fascination with any particular thing passes. Although even when his attention is drawn to something else, his original interests are still there, an accumulating set of old beloveds.

Well, she isn't willing to go in and out of focus in his life. And she certainly isn't going to become an old beloved.

But this is ridiculous. Surely this is about the fire, this thing—this disaster—that has been so peculiarly destabilizing, so disorienting.

And now the revelation about his junior—another bad blow. Jeremy the Traitor.

I thought I knew who he was, says Eduardo, bewildered.

But did you know *what* he was? thinks Geneviève. Misidentification of species. *Homo duplicitous.*

Interesting, the ability to be two-faced. An innate talent? A learned skill? At the very least, it would be tiring to keep up. And what did he think he was going to get out of it?

Hegge offered him a partnership, says Eduardo.

Maybe she should be thinking more fondly of the biology department.

Yes, that was bad, that was another blow for him—no doubt about it. But lately she feels her head is swelling with the urge to yell at him: *Pull yourself together. Snap out of it. Get some help. We have procreation to attend to. We have an organism to create. We have lives to live.*

Not things she could actually say to him—especially *get some help*, almost a foreign concept to him. His hands are healed now, although marked with white scars. But he was barely home during the race to finish the market design, coming in after she was asleep and leaving before she was awake. In that brief window when her chorus line of haploids was (at least theoretically) ready to dance, she had enlisted him in the cause one night, but it was such a strange, disconnected experience, it was almost vaporous.

Paroxysmi susurrus. A whispering orgasm. It starts with a faint fluttering that expands out in a moon-shaped, rippling motion, building gradually, haltingly, in dry ruffles to a sighing, soft smack.

Be patient. (But how patient?) Be sympathetic. (For how long?)

Don't be so patient, says Patrick.

Go away, says Geneviève.

A trick, something that Luc and Raymond played when they were young, floats into her head. In the evening, when dusk was settling in, they would position themselves on each side of a small street. They would stand opposite each other, holding their arms out, as if they were holding an imaginary rope across the street. When a car approached, they pulled abruptly on the invisible rope, shouting warnings to the driver. When the car squealed to a stop, they ran down an alley, leaping over cracks and potholes until they were safe. Then they would double over, out of breath, full of glee, in the half-gloom of the evening.

Matty, she thinks, alarmed, sitting up suddenly. Then she spots him—he has climbed to the top of a square headstone, and is standing there, arms outstretched above him in the sun, the temporary victor in his war with gravity.

In a small river, almost a wide stream, she is being carried along by the swirling current. The water is something alive, luminous. It feels like silk, like a physical substance made of light, on her skin. The effect is almost erotic, transcendent. She is being swept along in it, turning, rolling, overcome by astonished gratitude.

Now Eduardo is swimming next to her. A poor swimmer, he is splashing around clumsily, swallowing water. Look what I found, he says, although he says it silently. He has a wet bundle that he is trying to carry while he swims. He shoves it into her arms.

The water begins ebbing away, until she is able to stand up in it. It keeps going down until it is at her shoulders, then her waist, and in minutes, down to her knees. She begins opening up the bundle, layers and layers of cloth. Inside the bundle is a child—is it Chloe? No, not Chloe, a baby—a baby with an old face, smiling at her. She is overwhelmed by this, filled with a sense of utter relief, as if all the hardened, twisted parts of herself have dissolved. She holds the baby close, swaying slowly back and forth.

After a minute, she notices something is wrong. One of the baby's hands looks smaller than it did before. She bends over to examine it more closely. At first, she can't see anything, but then she notices that his hand is withering. She locks her thumb and index finger around the tiny wrist in panic, trying to prevent the withering from extending up his arm. Beneath her fingers, though, she can feel his skin wrinkling, and then shrivelling, a feeling that makes her shudder.

She kneels down, distraught, and unwraps the baby again. His entire body is drying up. Now she is frantic, hugging him, rubbing his arms and legs, looking around wildly for help. Eduardo has disappeared.

Despite her efforts, the baby's skin is slowly turning into a dried husk, his head is turning black. She clutches him against her chest again, and wakes up in horror.

She is making soup, a Portuguese *caldo verde*—a meal in itself, the cookbook had said. An experiment—aren't they all? A Berg suite is playing, a new recording by a quartet she loves, while she takes the papery skin off the onions and then chops them up. The opening allegro is an atonal movement, almost harsh—she is wondering whether they should include this in their repertoire, but she knows the first violin will object. He is probably right—they do have to have a mix, and the more melodic pieces are essential. Not only for the audience—these are the pieces, Haydn, the Bach, that clear away the debris between the four of them, that allow them to emerge on the other side cleansed and calmed, capable of new tolerances.

She begins sautéing the onions, and then forces garlic through the press. The music is beginning to build uneasily, while she peels and slices potatoes, then adds them with the stock. By the time she is slicing up the fatty chorizo, the andante has started.

Now the kale—she looks at it dubiously. A rubbery vegetable. But she rolls it up and cuts it into shreds, then adds that as well, listening as the music enters a high, breakneck passage. She stirs the soup, and sets out bowls, bread, a bottle of hot piri piri sauce, on the table.

A few minutes later the soup is done, and the suite is moving eerily into the third movement. The sounds are almost stalking each other, the notes tumbling and skittering around. Whenever the music starts to verge on conventional tones and rhythms, it swerves away again into the wild, the dissonant.

'What *is* this?' says Eduardo in English, as she ladles soup into the bowl in front of him.

'It's Berg's Lyric Suite,' she says in French.

'It's not my idea of lyric.'

What is your idea of lyric? she almost says. Instead, she says:

'What do you think of the soup? Does it have an aftertaste? Is there too much garlic?'

She reaches for the piri piri sauce and puts a few drops in her bowl.

'It tastes fine to me,' he says, absently.

They eat in silence for a minute.

'We were talking about difference tones today,' she says. 'At the rehearsal.'

'What?'

'Difference tones. It's when two pure musical tones sound together at the same time in a particular way. You can hear a third tone, not a blend of the other two, but an entirely new tone.'

It doesn't really exist, the first violin says. It's an auditory illusion.

'Interesting,' says Eduardo, although he doesn't seem particularly interested.

She adds more hot sauce to her soup, and takes a spoonful. Her mouth starts to prickle, and then to burn, and she swallows.

'It's only audible by ear. It doesn't register on scientific instruments.'

If you can hear it, it exists, the cellist says, standing his ground for once.

She picks up the hot sauce bottle again and shakes it over her bowl. The next mouthful sears her tongue, the roof of her mouth, the back of her nose.

Eduardo is watching her now.

'They can do a mathematical calculation to get the frequency. But that's only math, it doesn't really prove anything, does it?'

'I don't know,' he says, more warily.

She adds sauce to her soup again, and takes another spoonful. It explodes in her mouth, and she begins to perspire, her eyes start to water.

They can't measure it because it doesn't exist, the first violin says.

'Some people think that it's actually produced inside the inner ear, a kind of distortion,' she says.

She can see Eduardo hesitating, unsure of what to say. He wipes his mouth with a napkin.

'What do you think?' she says.

'I don't know. I'm not the musician here,' he says, sitting back in his chair.

'But what do you *think*?' she says.

She holds the bottle upside down, and watches the hot sauce drip steadily into her bowl. Then she picks up her spoon, and waits with it poised in her fingers.

'I suppose it's possible,' he says reluctantly.

I've been having bad dreams, says Geneviève.

I don't remember my dreams, says Eduardo.

Dreams? says Patrick. If I had dreams, it would mean I was sleeping.

She is sitting in the dark with Patrick. The original plan was for the three of them to go to the play, to get a babysitter (a babysitter for the babysitters, she says) but Eduardo dropped out.

I'm still exhausted, he said. I'm getting too old to do these work marathons any more.

I think I'll go anyway, she had said casually. Matty's already asleep.

Fine, he said.

She should be watching the stage now, but her mind keeps wandering, despite the raw poetry, the energy of the play. Maybe because the story is a bleak one, and she has to keep biting her tongue not to make sardonic remarks. Or maybe because she is acutely aware of Patrick sitting next to her, the rough fabric of his jacket sleeve, the shape of his head. It is all she can do not to rub her head against his shoulder, not to leave it there and close her eyes.

She has to admit that the staging is impressive, even if the play is grim. The script is a series of interlocking monologues, and images and pieces of scenery slide onto the stage from every angle, silently overlapping, receding to reveal other pieces, dividing and then disappearing. This seems to be a combination of ingenuity, image projection, and lighting, at least as far as she can tell.

She is watching and listening to the words without concentrating much on the meaning—the virtuosity of the production isn't entirely a positive thing. Occasionally it seems to be competing with the play itself, with the power of the script. Most of the time, though, the performance and the text of the play flow together, each heightening the intensity and illusion of the other.

The climax of the play is approaching, a tragic twist. She glances at Patrick, and he turns and smiles at her, his eyes warm. She looks away quickly.

Afterwards, they walk along Saint-Laurent, peering in lighted store windows, watching the people on the cold sidewalks. The crowds seem to be unusually excited and hopeful, as if they were expecting something delightful or satisfying—as if they were all on their way to parties that might be full of vague things they had always wanted, but couldn't quite envision. Most of them are carrying something to drink, metallic water bottles, long cans of soda, coffee cups with white hats. Maybe all this hope was thirsty work.

They pass by a construction hoarding, covered in posters and graffiti. Out of the corner of her eye, she sees a scrawl in black spray paint.

Anglos, go home.

The E of 'home' trails down in a drip. Below it is another scrawl in red paint.

We are home, thank you.

'Isn't there some kind of statute of limitations on historical sins?' says Patrick in French. 'Is this ever going to be over?'

He says this casually, although there is something else in his voice.

'Over?' she says. 'You mean like having the measles? Or did you want a precise date—say 2:30 on June 23, 2023?'

'A date would be nice—although I was thinking more like next Tuesday.'

He must have caught some of the crowd's hopefulness, she thinks.

Ahead of them in the street, a man in a red vest with a yellow X is sitting half in and half out of an open manhole, surrounded by orange pylons. Beside him, an air compressor on wheels is rumbling. The man is peering down the hole, speaking to someone below.

The cars have been channelled into the other lanes, and they are moving slowly. An impatient driver is attempting to edge around the pylons.

'Let's cross,' she says.

As they set out across the street, the edging car comes too close to the manhole, knocking over several of the pylons.

One of them falls on the man in the vest, who recoils.

'*Maudit Christ,*' he yells at the back of the car. '*Qu'est tu fais là*?'

Patrick picks up another one of the fallen pylons, and sets it up again. The man in the vest nods at him.

'There must be some potential for a lawsuit there,' says Geneviève. 'A near miss. Mental anguish.'

'He didn't seem very anguished,' says Patrick. 'But I don't do that kind of law, anyway. Although I seem to be saying that a lot recently. It's amazing how many kinds of law I don't do.'

'Let's not go home yet,' she says suddenly. 'Let's have a drink. Let's get out of this cold.'

'Of course,' says Patrick. 'We certainly deserve it. In fact, it should be required after a play like that—it seems designed to make you want to throw yourself out a window.

I don't know which was worse—the psychopathic wife or the suicidal lover. My guess is the playwright's home life leaves something to be desired.'

'Why do you think it was autobiographical? He probably has a very cheerful and placid home life.'

'I doubt it. Although I suppose anything's possible—maybe he became a playwright to explore his dark side. In which case, frankly, he's been a little too successful.'

They pick a bar randomly, a place with red lacquered tables, black chairs, cocktails with ingredients that seem to have wandered in from another menu—cardamom, tobacco, coriander.

'I can feel my sophistication quotient rising by the minute,' says Patrick. 'Although I think I'll stick with Scotch, something I actually like.'

She orders an exotic drink, fascinated by the ingredients—scientific inquiry, she says—but it tastes like perfumed wood when it comes. She gulps it down quickly to avoid the taste, and orders something else.

'That was impressive,' says Patrick. 'I see I have my work cut out for me if I'm going to drink you under the table.'

'Not really,' she admits. 'I can get it down, but I can't hold it very well.'

'You just need more practice,' he says. 'I could be your coach. Put yourself in my hands, and in a few short weeks, I could make you a world-class drinker.'

'Are you a world-class drinker?'

'Of course not, that's why I would be the coach instead of the contender. But I've certainly had the chance to study world-class drinkers.'

'Like who?'

'Oh,' he says, suddenly vague. 'Relatives, friends. It's a surprisingly common talent.'

'Not so surprising,' she says wryly.

Keep talking, she says to him silently. Keep talking, so I don't think about your mouth, the shadows under your eyes, the side of your jaw.

Everything about him seems to have become more pronounced, more defined. And she feels as if her senses have become cross-wired—looking at him, she can almost feel the texture of his skin.

But he does keep on talking—obligingly? nervously? habit?—a swirling tide of words that seems to surround her, to flow into all the spaces inside her, to fill all the little cracks and grooves. She thinks of herself as an effortless talker, but compared to him? How can he possibly manufacture all these words, where does he get such facility, such dexterity? Some school for advanced fluency? He must have been their top honours student—their valedictorian.

'But I'm doing all the talking here,' he says. 'Is it getting too late? Do you want to go home?'

'No,' she says. 'I don't.'

'Why don't we start walking again?'

So they do, drifting down side streets, their footsteps muffled by the dead leaves trampled on the sidewalks.

'Do you mind if we stop in at my office?' he says. 'I forgot a file.'

'Of course not,' she says, trying to shake off the sense of being in some magnetic current, being pulled along in his wake.

His office is in a converted house, limestone on the outside, elaborate rugs on the inside, Kandinsky prints on the walls.

'Poor Kandinsky,' she says in the hall. 'Good thing he isn't alive to see his work turned into universal office décor.'

'Except he would be rich,' says Patrick, opening an office door, turning on the light. 'That's always a consolation. Often extremely consoling, I understand.'

'He had some theory about the correspondence of art and music—that hue was like pitch, that saturation was related to volume, something like that.'

'Sounds impressive.' He is rummaging through the stacks of files on his desk, checking labels.

'It would probably sound more impressive if I could remember what it was.'

'Got you,' he says to a file.

He wrestles it into a manila envelope, gives a last look around and heads to the door, where Geneviève is waiting. He switches off the light, putting an arm around her shoulders to guide her out of the room. She turns to him, confused, and instead of turning away, he pulls her in, pulls her close, their bodies pressing together. Then he is running defiant, hands through her hair, kissing her face, her ear, her mouth.

This is a mistake, she tries to say, but her tongue is too thick, and she is flattening herself against him, lifting up her arms to his neck as his hands slide down her body. This is a mistake, she says silently, as ancient layers inside her unfold and recurl, swelling out from each other. Then she is twisting her body, and they are down on the floor, pulling at each other's clothes urgently. This is a mistake, she thinks, her hands on his hot skin, the sensation of it against her palms. This is a mistake, she thinks as she feels her body rising in a strange new rhythm, a rhythm of aching, longing, pleasure so intense that it swallows up everything else.

Paroxysmi maerentis. A bittersweet orgasm. It starts with a rush, building in wild, rough waves, singing from height to height, dark green and muscular, until it reaches a searing, shivery crest—and then crashes down, leaving only traces of lush, briny melancholy.

She is distracted during her class, muzzy-headed, losing her train of thought several times. The students laugh good-naturedly, curiously, the second time it happens, and she feels a ripple of embarrassment. They are suddenly paying more attention, alert with the finely honed student instinct for professorial weakness. This would be a moment to tell them something particularly weighty, particularly elemental—some organizing principle for everything. But what would that be? Her peroxisomes and ribosomes have deserted her.

I have touched his bare skin, she thinks, instead of *who can describe the characteristics of water conducting cells?*

Epidermis, dermis, hypodermis, she thinks, instead of *what are the elements of cohesion-tension theory?*

He has had his tongue, his hands on me, she thinks, instead of *what is the process of transpirational pull*?

Was it what you expected? said Patrick.

Of course.

Not in a million years.

She gives the students an assignment—how plant hormones respond to light, touch and injury—and lets them out early, watching them pour out of the blue-panelled lecture hall, talking animatedly at the thought of a few extra minutes to themselves.

A student who wants to appeal his grade looms up, shorn reddish hair, pale eyebrows and lips, as if his face had been bleached. She tells him to make an appointment and flees to her office, closing the door and then leaning up against it for a minute.

Her brain is a hiccupping muddle of elation and wretchedness, chasing each other around in an endless circle. She feels as if she has become a strangely unfamiliar version of herself, like a new arrangement of a piece of music.

Is this all it takes? she thinks. Do I really live such a relative life, so defined by bouncing off other people, so shaped by encounter? Her personhood so reliant on a particular configuration of people around her that all it takes is a new human reference point to turn her into someone so different? Or is it just the exhilarating discovery of being newly wanted? Or the relief that happiness like this, happiness of this giddy strength is still possible, even momentarily?

But when she thinks about Eduardo, she is clutched with guilt so corrosive it feels as if her veins are being stripped.

I can't give it up, she thinks in a panic.

I can't go on with it, she thinks with despair.

No need to decide right now.

This must be stopped right now.

She wishes there was a way to map the geography of this—this event?—or its anatomy, its physiology, some way to consider it in orderly, analytical terms. But would that help? Probably not. This is all in the realm of the marshy, the chaotic, the inexplicable.

On the way home, she picks up Matty from Luc's girlfriend, who almost pushes him out the door.

'He's running us ragged,' she says in French, one sobbing little girl in her arms, another calling from another room.

Matty is undaunted, chattering as they walk holding hands, pulling her arm, then jumping on to it to swing for a second.

At home she finds herself overwhelmed with gentleness towards Eduardo, but a bloodless gentleness, as if he were her patient and her task was to treat him as delicately and kindly as possible. What does this mean? she thinks, as if it were a clue of some kind, as if she could track down her own feelings by external observation. But this seems as opaque as anything else.

She is putting Matty to bed—*Asterix* this time—when there is a flurry of knocking at the door. She hears Eduardo get to his feet stiffly, the door opening, a gabble of voices, angry, indignant. Familiar, though, not hostile.

'Back in a minute,' she says to Matty, but he refuses to let her go, protesting loudly, suddenly clinging. She strains her ears to hear something through the closed bedroom door, but he is climbing over her, looking for a more comfortable spot in the hollows of her body, in the curve of her arm. When he finally falls asleep—mid-wiggle—she extricates herself, rolling him gently into the bed and pulling up the quilted coverlet around him.

In the living room, Eduardo is sitting with a senior associate, and two juniors. They turn towards her, arranged in varying postures of dejection, like some odd tableau of defeat.

'What's wrong?' she says, alarmed.

'There's been a leak in the market competition,' says Eduardo in a strangled way. 'They're going with the Hegge design.'

Chapter Twelve

Patrick is lying on a portable bed, watching his blood scurry down a tube into a plastic sac. As the flow becomes continuous, the nurse, satisfied, flicks the bag with her fingers, and then moves on to the next cot. His blood looks dark in the tube—like red wine, he thinks, or pomegranate juice, not like the bright splash of colour when he accidentally cuts himself.

He is a sporadic donor—sometimes curious about the people with carefully filled out record cards, every two months to the day. The Good Citizens of Blood. He is a once-in-a-while person, who walks by a clinic in the lobby of a building—beds trundled in, cloth baffles positioned—and thinks: why not?

But first the questionnaire.

Have you ever had malaria? No.

Have you ever had hepatitis? No.

Have you been in prison during the last twelve months? No.

Have you ever had a job handling monkeys?

'Do many people say yes to that?' he says. The volunteer—a small man who seems to be losing a battle with his large sweater, giving him a porcupine-like quality—ignores this and looks up, pen poised.

No.

Have you been treated for syphilis or gonorrhea? No.

Have you taken money or drugs for sex? No.

Apparently some would-be donors lead more interesting lives than he does.

He is aware that this donation—today, anyway—could be considered a faintly ridiculous act of penance. But he doesn't feel penitent. Light-headed, yes—and last night he was exuberant. Is there anything headier than this? Even now he feels as if his brain was carbonated—not necessarily in the most pleasant way, but consisting of sharp, disconnected particles. But guilty? No. Remorseful? No. This is not some black and white puppet show. This is infinitely complicated, infinitely enigmatic reality. No, he refuses to be hijacked by some outdated taboo.

The needle in his arm slips slightly under the tape, pinching his skin painfully, and he shifts his arm a little.

'Feeling all right?' says the porcupine-like man briskly.

He nods—truthfully.

He is a little surprised himself that he doesn't feel more conscience-stricken. He has managed to distance himself from his original self-accusations—disloyalty, betrayal—words that sound dusty and medieval. Leaving aside the usual conventions—which he does—this is still a deception of sorts. But when he probes this, tries to locate any residue of guilt, he feels only blankness. Of course, he could tell Eduardo about it—undeceiving him—but this seems completely absurd. It would do Eduardo no good—an exorcism of his own conscience at the expense of Eduardo's state of mind. State of mind—hell, sanity. Eduardo's faith in people is already so badly shaken, his mood already so dark and brittle—no, it would be disastrous, might send him over the edge. Unless he is over the edge already.

For the first time, it occurs to him that he has been assuming Eduardo is going through a temporary rough period. What if this isn't true? What if this is a downward spiral leading to a permanently altered life? He thinks of a man he knew in law school, an alcoholic who now lives in a basement apartment with thin broadloom and a mildewed shower curtain, and who calls him every so often for a loan that will never be repaid.

No, Eduardo is more resilient—more arrogant—than that. Isn't he? But either way, what would be the point of him knowing about this thing with Geneviève? Nothing. It would only add to his afflictions and for no good reason. It would be utterly pointless. And who knows what this is at the moment? Who knows what it will turn into? No, at the very least, telling him would be premature.

He lifts his head to look down at the plastic sac of blood, rocking in a mechanical cradle. Half-full.

Giving blood makes him feel unavoidably virtuous. The thought of being paid for it—as they do elsewhere—seems barbaric, like trafficking in body parts. As it stands, this is one thing he can do which is entirely, incontrovertibly, good. Whatever that means. Although there is really no definition of *good* that would not include this. He knows there is something banal about this thought, but not everyone can be good in interesting ways. And maybe the fundamental nature of this act, the utter goodness of it, trumps its banality, its sentimentality. Well, perhaps this is getting too carried away.

Although isn't *carried away* what he wants? Carried away from the erosions of routine, kicking out the ossified structures on which everyday existence is hung. This is something he understands—more than understands—about Eduardo, his restlessness, his hunger for newness. Does this mean Eduardo would understand about Geneviève? No. There is some old world seam running through Eduardo, something distinctly unmodern. And why take the risk, anyway?

He knows this is rationalizing—a talent honed in law school, the veritable Olympics of rationalization. Pick the right premise, and it will take you anywhere. He doesn't believe in the kind of regulated reasoning that can be insulated from subjectivity. But he has to think about Geneviève, this thing, in some way, in some form or another, and this is the way he thinks. The fact that these arguments lead to a convenient conclusion doesn't necessarily mean they are unreasonable or unsound. Suspect, maybe, but still potentially sound. Unless he is rationalizing his rationalizing.

In a different time and place, Eduardo could have sued him for damages. Criminal conversation—this is the old English tort, an antiquated term. For this, or for enticement—something that sounds closer to the mark. Had he enticed Geneviève? Had she enticed him?

But if these old rules are useless, is it merely a moral free for all? Of course not. First, do no harm. Not his profession, but a sturdy, all-purpose proposition. Who are they harming? Not Eduardo, if he isn't aware of it. Is there some more ephemeral harm regardless, an insult to his essential personhood, whether he knows or not? Only if it *is* an insult—circular reasoning.

But surely this will damage his friendship with Eduardo, the deception turning into a form of dry rot? Well, maybe *do no harm* is too all-purpose, too unnuanced, then. There is always harm.

He needs some way to put this to the test—some kind of experiment entirely aside from these looping thought rings. He could drop by Eduardo's office casually—he has another marked-up contract to give him—and see if he can face him. Not to tell him anything, of course—but only to see if he can silently sustain his views in the presence of Eduardo himself. A kind of moral assay.

But what will this really show? If he still feels so unrepentant, still so unmoved, does that make him any more right? Or simply more unfeeling? Wouldn't this also be pointless?

And would Eduardo detect anything, sense some change in him? Has he already sensed some change in Geneviève? Something amorphous, a sense of redefinition, a quiet galvanization? Probably not.

You think I'm oblivious? he says to Beth. Let me introduce you to the king of oblivion, the prince of not noticing.

But is it possible Geneviève will tell him? A horrible thought—he almost jerks himself up into a sitting position, and the needle twists again.

'Keep still, please,' says the porcupine-like man walking by, his voice issuing from the depths of his sweater.

But surely she has reached the same conclusion? Difficult to know—he will have to talk to her. He feels strangely light-headed about this. A good excuse to see her again, to see how different things are, how rescrambled they are. Now that he has stroked the hollows in her neck, the soft creases of flesh between her legs.

Of course, things might not be that different.

Does he want them to be different?

He looks down at the rocking bag again. Almost full. But suddenly, the sight makes him weary.

He closes his eyes—and then opens them quickly. Dangerous, lying down like this.

We have your results, his doctor had said. They've eliminated sleep apnea, REM behaviour disorder, periodic movement disorder, hypnologic hallucinations, narcolepsy, and sleep paralysis.

What's left? said Patrick.

Non-specific insomnia, said his doctor, pronouncing this in a way that had something reproachful about it, as if Patrick were at fault for not having a more definable disorder.

Is there anything I can do about it?

The same things he has already tried over and over, hoping that they will suddenly start working. Surely, he thinks, in a country where almost everything can be bought and sold, he should be able to purchase sleep—not sleeping pills, but actual sleep. A country that has produced a bed in a bag should be able to produce sleep in a handy, resealable package.

You're lucky, the doctor said. My next patient is on dialysis.

Lucky. Patrick knows this. He is lucky, lucky, lucky. Look at all that luck.

He will have plenty of time to contemplate this. Particularly at night.

The November day is cold, but a bright, dry wind is gusting through the streets, shaking the branches of trees, tossing flyers into the air, sending plastic bags sailing down the sidewalk.

They are sitting in a park, some of the trees still disfigured by a long ago ice storm. The wading pool in front of them is empty, closed for the season, dead leaves scattered around the shallow bowl. Despite the chill, two boys are playing on a tennis court nearby, knocking the ball around in a desultory way.

'What difference does it make to us?' says Patrick in French. 'I don't mean to be callous—I'm sorry that he didn't win the competition, but he's lost competitions before. And won them. So what does that have to do with us? It's irrelevant.'

She is pale, agitated, her dark hair spread out untidily on the shoulders of her red jacket.

He moves to take her hand, and she flinches away.

'This was an act of insanity, anyway,' she says. 'For both of us,' she adds hastily. 'We can't do this, make the same mistake again.'

Of course she would have to say this.

'No,' he says. 'It wasn't insane, it wasn't insane at all. I think there must have been a number of different reasons, undercurrents, strands of things that were cumulative, ingredients that suddenly coalesced. But it wasn't insane.'

Is this really what he thinks? Or is his tongue merely getting away from him?

'I'm not sure that makes it any better,' she says slowly. 'And it certainly doesn't mean we should keep going.'

'But all the elements that fed into it are still there. We have to have some faith in our instincts, that they represent something.'

'Really?' she says wryly. 'I think we might have an inkling of what they represent.'

'No,' he says. 'That's where you're wrong, it's not that simple. You think we're the kind of people that would just give in to impulse, without there being something more profound behind it?'

'I don't know *what* kind of person I am any more,' she says, twisting her hands. 'I barely recognize myself, let alone you.'

'But that's the point. This is something that allows us to reshape ourselves, to reconstruct who we are.'

'But if we wanted to do that, why do we need to do it in relation to someone else—why can't we do it ourselves, on our own? It's as if we were hitching a ride on someone else.'

'I don't know,' he admits. 'But it's not just recasting ourselves—it's the new thing we generate between us, a new combination, like a new alloy.'

Is this nonsense? he wonders. Or has he actually hit on something here, sounding out a real thought, pulling something clear out of this haze of feeling?

Longing flickers across her face for a second.

Then she says: 'No. No, it's impossible. Especially now, with Eduardo losing the competition. He's just so—wounded.'

'Yes, but that's not your fault. You're not responsible for that. Why should it stop you from doing something that gives you this kind of chance? What if you never come across this again?'

She says nothing.

He tries to take one of her hands again, and this time she lets him. He traces the outlines of her fingers with one of his, sliding into the pockets between each finger. She is still now, and she watches him do it mutely.

He can hear the rhythmic thwock of the tennis ball on the court behind them. The wind chases the dead leaves around one side of the wading pool, and then retreats.

'No,' she says again softly. 'We just have to erase it all from our minds, delete it from our memories, from everything.' But she doesn't pull her hand away.

'You must have a pretty obliging memory if you can delete something like that,' he says lightly. Her hand is warm.

'But what are you suggesting?' she bursts out. 'That I leave Eduardo?'

'No, no,' he says quickly. 'Nothing as drastic as that. That's why it's benign, why it doesn't hurt him. As long as he doesn't know. No, all we would do is continue to see each other, be together like the other night.'

He says this as if it were something mild, innocuous—who could object to such a thing? But as the words come out, he is conscious that this sounds weak, almost anemic, even to his own ears.

A flock of small black birds rises over them suddenly, as if someone had tossed them into the sky. They fan out and begin wheeling around in loops against the ribbed clouds.

'You're probably right,' she says.

He feels a surge of relief mixed with triumph.

'But it still seems mean-spirited to Eduardo, whether he knows or not. And I can't imagine dividing myself into the two people necessary to carry this off. I can barely handle one of me.'

She gives a strained laugh.

'I don't think I'm someone who's good at being bad.'

'Being bad?' he exclaims. 'Being bad? Is that how you think about it? In such small-minded terms?'

'Yes and no,' she says.

'I think you're afraid,' he says. 'You're dressing this up in moral terms, but you're afraid.'

She looks at him uncertainly, hesitating.

'Think about it,' he says.

'For a change,' she says drily.

She stands up, and touches the side of his face with the warm palm of her hand. Then she hesitates, almost says something, and starts to walk away quickly.

Wait, he wants to shout, to run after her, to call her back. Wait, this can't possibly be the way things are left. This is not the right conclusion, not the right outcome. You've confused this with some other situation, some other person. I'm not even a risk-taker. When did I turn into the risk?

But this ball of words is stuck somewhere in his chest, he can't get it out past his throat and onto his tongue to be launched towards her. All he can do is watch her red jacket getting smaller and smaller in the distance.

She didn't say no, he reminds himself. All she needs is a little time to think about it. Then she'll see the wisdom of this. There were all the microscopic signs of attraction, the fact that she was clearly fighting against herself. She needs some time to get used to the idea, she is panicking at the moment. And the loss of the Eduardo's design competition will fade, she will feel freer. He is buoyed by this thought, increasingly convinced that she will come around.

By mid-afternoon, he thinks: she didn't say yes, either. He thinks of her hesitation, her flinching—all the microscopic signs of withdrawal, of retreating. Maybe the magnetic pull of that night will fade instead. *Erase it from our memories*. His optimism begins to recede.

By the evening, he thinks: she is not going to come around. He has been deluding himself. This affair—if such an embryonic thing could be considered an affair—is finished. If there is any hope left, it is a tiny, flabby hope.

He is torn between anger and astonishment. If there was one thing that he hadn't predicted for his new, shiny, post-divorce life, it was the role of luckless lover.

Would it have made any difference if he had said: *leave Eduardo*?

Some things go on relentlessly, and one of them is Chloe's pick-up schedule. Normally he would brace himself for

Beth's wall of animosity, but he is already immersed in a kind of dreary limbo—convinced of his failure with Geneviève, but not quite convinced enough—and this gives him a certain protective coating. Go ahead, do your worst, he thinks.

She looks at him curiously, though, standing in the hallway by the door, holding Chloe's backpack.

At least Chloe is glad to see him, patting his face. He scoops her up from the doorway, and then dips his arms so her head is lower than her legs, and she gurgles in delight.

Beth's expression softens a little, and unreasonably, he feels a tiny easing.

Then her face closes down again, she hands him the backpack and shuts the door.

At home, he gives Chloe dinner on the coffee table in the living room—half a jam sandwich and some grapes is all she will eat. He is trying to teach her the constellations, so while she eats, he tells her about the November night sky—Cynus, Camelopardalis, Pegasus.

'A winged horse,' he says, flapping his arms. She flaps her arms obligingly several times, puzzled. He realizes that she probably doesn't know what horses are, or that they normally don't have wings. What a shapeless two-year-old world she lives in at the moment—so few fixed points. Or maybe more accurately, a handful of familiar things—her toys, her bed, her bottle, Beth, him—surrounded by a world in which everything else is an unknown possibility. In which things might have any powers or characteristics at all. A bar of soap might be able to talk, a chair might do acrobatics, a stove might sprout trees. As all these things settle down, become fixed, will she feel this as a loss of elasticity, a loss of enchantment? Or a gain in stability? Both?

Tired of eating and flapping, she crawls up into his lap, and begins sucking her thumb. He finds this oddly comforting. At least there is Chloe, he thinks. At least for now. If he has managed to turn everything else into a disaster, at least there is this small, living, breathing entity, this little thumb-sucker.

He is angry at Geneviève, but his anger has congealed into a solid mass. He has enumerated all the ways in which she is wrong, honing the same thoughts over and over. Unless she comes around, in which case she is right. But no, this is still so unlikely. And in this unlikeliness, he has made his case against her—her irresolution, her lack of fortitude. But even he is dimly aware that there is really only one crime to her discredit—not wanting what he wanted. Or not wanting it enough.

It doesn't help that he can't even win these arguments in his own head.

For instance:

How could you go so far and then retreat? says Patrick.

Sometimes things become clearer as they go along, says Geneviève.

How can you think of giving up something so gratifying, so tantalizing?

I'm not sure it's worth the cost.

How could it not be worth it?

Is that a real question? she says.

Or:

How could you let me fall for you like that? says Patrick.

You did that to yourself. You're an adult.

Yes, but you led me on, encouraged me.

Caveat emptor, she says.

Sometimes he tries to explain it to her clearly, neutrally, without emotion. As a practical matter.

This is the way it works, he says. When we meet certain people—people we have an affinity for—we edit and reshape ourselves to fit them, to make ourselves into the people they need or want. Only to a limited extent, of course. We merely adapt ourselves, we intensify or muffle certain features, we develop or discover others. For example: you needed me, and I remade myself into the person you needed. Now it's your turn. You see how it works? Really, it's not that difficult. A little flexibility—that's all. Call it a doctrine—the doctrine of reciprocal transformation.

Who *are* you? says Geneviève.

Chloe is sleeping now—she falls asleep so instantly, so casually that he is almost jealous.

He should wake her up—this is too early for her to go to bed—but it seems unkind. Besides, he finds her sleeping calming, as if she has scattered her careless abundance of sleep around her, as if he is the beneficiary of second-hand sleep. So instead he sits there, listening to her breathing, watching the early evening light stretch itself out across the room, turn reddish-gold, and then gradually fade into mauve shadows.

As he watches the light disappear, he is struck by the thought that it might be possible to take a more deliberate hand in his own decisions. Following the dictates of some subterranean forces in his psyche seems to be working out badly. So far, these conspirators seem to have been remarkably incompetent. And now the Patrick who was insulated by irony, the professional bystander, the adept avoider of conflicts, seems to have deserted him as well—drowned first in infatuation, then in the loss of it.

But maybe it is possible for him to be more deliberate, exercise more agency. This small, ordinary thought seizes him, seems to unfurl and spread out, become something remarkable, something occupying every part of his mind. Why not? He could hardly do worse.

Perhaps he should wrest control—or something like it—from these conspirators. Maybe he should be taking matters into his own hands. Maybe instead of stumbling in and out of other people's orbits, he should pull all the disparate parts of himself closer together, and see if they can agree on something once in a while.

Chapter Thirteen

What a fool I am, thinks Eduardo.

He rakes his fingers through his hair, staring blankly down at his desk.

How could he possibly think a more risky design would work? What possessed him to take such a chance? The first design was perfectly fine—better than fine.

But perhaps it didn't have anything to do with the design—maybe the market owner had simply wanted Hegge on the project.

'I'm no good at this,' he says out loud.

He sees a wisp of smoke drifting by, then several more, and he grips his head more tightly.

The designing, yes, he can do that. Or at least he feels he has a fighting chance there. But the politics—the courting of clients, the glad-handing, the maneuvering—these things elude him.

I don't have the necessary cunning, he thinks dismally.

Although this is not only a matter of ability—it also requires some faith that a particular strategy will work, the other people will react in ways that are consistent with the plan. People—and events—seem far too unpredictable to him for this. Unamenable, even inimical to carefully laid plans. Or even carelessly laid plans. The whole thing a peculiar game of chance—certainly not a game of skill, or not a game of skill that he knows how to play.

More wisps of smoke are gathering, along with an unpleasant burning smell. He wonders vaguely if someone has been using the office microwave.

Maybe he *is* too stiff-necked, as Patrick said. Maybe the real problem is that he finds all this distasteful, that he is unwilling to engage in these degraded forms of social interaction. Maybe he is secretly complacent about his failure to learn these skills, as if this showed that he was a person of higher character.

The reality is that these are scruples he can't afford.

Are you sure that you can afford *not* to have those scruples? says Álvares.

I am now, says Eduardo.

Although perhaps the problem was losing his temper at the concert. Undoubtedly Hegge had told everyone he knew, made sure that this story made the rounds. Who wants an out-of-control architect? He almost groans out loud. What is happening to him? How could he have lost his temper like that?

Does this mean his own reputation will be unfairly tarnished forever? First by the arson allegations, now by this moment of madness. The thought gives him an almost physical pain in his throat.

What are you, a child? he says to himself. Life is still nasty and brutish, even if it isn't quite as short as it used to be.

'Two hours,' his father says in Portuguese. 'She has never been gone this long before.'

'Has anyone seen her?' says Eduardo, standing in the doorway. The sky is overhung with dark grey, a cloud mass that has crowded out the afternoon sun.

'Not for a while.'

'Where does she usually go?'

'Down the street a block or two,' says his father.

'How can she be lost?' says Eduardo reasonably. 'She knows this neighbourhood so well. She's probably in an alley, straightening up someone's garbage cans.'

'You don't understand,' says his father.

Eduardo looks at him sharply. His father has an odd expression on his face, a mixture of worry and something almost furtive.

'She's—not herself.'

'I know that,' says Eduardo impatiently.

His father looks away.

'You mean it's gotten to the point where she might not even know where she is?' he says.

His father is still looking away.

'Why didn't you tell me?' Eduardo says angrily.

'I don't have to answer to you,' says his father, his back still to Eduardo

He is ashamed of her, Eduardo thinks. The bleakness of this almost overwhelms him for a minute.

'Where have you looked?' he says to his father grimly.

'All along Deluth, and a few blocks north and south,' his father says. 'But she could have been on one street while I was on another.'

Two hours is a long time to wander. Although she might have been wandering in circles. He has a picture of his mother, her short, swollen body hidden by her black coat, standing on her bowed legs at some busy street corner, bewildered, fearful. Saying hopefully to the people walking by: *spik Portuguese? spik Portuguese?*

'You go towards Sherbooke, I'll go towards Saint-Denis,' he says.

He walks urgently—almost jogging—stopping to peer down laneways, behind houses. He passes a doorway with a tall black rooster painted on it, a high red coxcomb like a pompadour, and clusters of stylized red dots and hearts on its black body. *Bom Dia!* says a cartoon bubble coming out of its beak.

More clouds are moving in sluggishly, and the air has become expectant. He stops and blows on his cold hands for a minute. All we need is rain, he thinks. A grumble of thunder sounds in the distance, and he wonders if he should go back and get his car. He can cover more ground that way, but not as thoroughly, and he will be limited by the one-way streets. He glances up at the dark sky, but decides to keep on going.

He ducks down a lane, but all he sees is a woman at the back of a house, taking in laundry from a clothesline. She plucks off the clothespins swiftly, expertly, yanking the line towards her on its pulley.

Then he is at rue Saint-Denis. Would his mother venture this far, or turn back? He looks down the street, where a shopkeeper is taking a sidewalk sign inside. The stream of people on the sidewalk is moving briskly, purposefully—no-one is browsing idly under this sky.

Eduardo hears the thunder again, and curses.

No, this isn't her kind of street—too busy, too intimidating for her. He decides to turn back and go north, following the narrower streets.

The buildings are so close to the curb here, he thinks that people must fall out of their doorways into the street in the mornings. He passes a *churrasqueira,* where a row of glistening brown chickens is rotating slowly on a spit, and then walks through the blast of roasted meat smell coming from a vent.

By the time he arrives at rue Marie-Anne, the street is strangely deserted, only a woman hurrying along, pushing a stroller, an elderly man already taking out an umbrella, a few others moving swiftly with bags on their arms.

He follows rue Marie-Anne for a few blocks, and then turns abruptly and heads down avenue Coloniale—he has no plan except to cover as many streets as he can. A sprinkle of cold rain starts, almost tentatively.

Three students are unloading a car in front of him, two carrying boxes of books into a basement apartment, the third standing indecisively by the open car trunk, looking up at the sky. Another rumble of thunder, and the third, too, grabs a box and hurries into the apartment.

The rain is stronger now, and he thinks of his mother out in the frigid downpour. At least she isn't bothered by thunder—or at least not up until now, he thinks, remembering. Who knows what might frighten her now? Although maybe there are advantages to a fading memory. Maybe old sorrows, old bitternesses disappear with it.

But the rain is coming down harder—he is already wet. What if she ends up with pneumonia?

Superstition, says Geneviève. Pneumonia is caused by a virus or bacteria.

But being wet and cold could lower someone's resistance, thinks Eduardo. Especially if that someone is elderly. Especially if she is out in the rain for a while.

He breaks into a jog again.

A lone bicyclist goes by in a yellow raincoat, head down, his bicycle seat covered with a plastic bag, a box perched precariously behind it.

Eduardo arrives back at Deluth, more alarmed. What if they can't find her? He thinks of newspaper stories about elderly women who have been robbed or assaulted, and then pushes them out of his mind angrily.

At the corner of rue de Bullion, he stops to take a few deep breaths. A gust of wind catches him, and drives the rain harder. As he straightens up, he glances at the Portuguese *épicerie* on the corner. Then he stops.

The store has turned on its lights, and he can see inside clearly. She is standing there, in her black shawl, filling a bag with oranges. She seems entirely at home, comfortable, raising her head to speak to the man behind the counter. He points, and she moves further down, takes another bag and begins turning over potatoes, checking for spots. While she does this, the shopkeeper takes a container of unripened cheese out of the cooler, then cuts her a length of *chouriço* sausage. The man says something to her, and she smiles, a glint of silver from a tooth, and says something back.

She moves over to a bin of white cornbread, hovers for a minute, then takes out a loaf and adds it to the pile on the counter. After this, she picks up a melon and puts it close to her ear, and then knocks gently on it.

Outside, Eduardo stands there, soaked and cold, and watches her. She knocks on the melon again, and listens to it intently, nodding her head slightly, as if there were nothing else in the world more important than this.

You misplaced your *mother*? says Geneviève. St. Antoine is really falling down on the job.

It could have been serious, says Eduardo sharply.

Sorry, she says, even though she isn't.

'I should get going,' she says the next day. He has agreed to stay home for the morning to look after Matty while she teaches a class.

'Fine,' says Eduardo, looking around for him.

'He's in the other room, in front of the television.'

'All right.'

'This is the last time—he's going home tomorrow,' she says.

'You told me.'

She looks at him a little dubiously.

'We'll be fine,' says Eduardo again. 'Go ahead.'

She leaves quickly.

He is relieved. She seems odd lately—flashes of warmth, often followed by a more searching look that makes him feel as if he is being catalogued. Or possibly measured for a suit he hasn't ordered and might not want. Could this be the fertility drugs?

Sometimes he wonders if all this—the drugs, the schedule, the cycle of anticipation and disappointment—are worth it. To her, obviously—and she has the worst of it. To him? He isn't sure. She has so entirely occupied this space that he hasn't really had to decide. He has certainly become accustomed to Matty—*fond* might be too strong, but he does seem very familiar now. And he is intrigued by the boy's perceptions, how he sees the world without the disfigurement of experience. A beginner of the *homo sapiens* variety. He himself is so far from this—what is the opposite of a beginner? An old hand?

Well, his hands are certainly older-looking, and not just because of the scars from the burns. He suspects he has a build-up of moral scar tissue as well—too much exposure to the absurdities of being. Particularly recently, when he seems to have been enrolled in a crash course in fallibility—not only his own, but that of other people as well.

But he will have to make the best of it. If they have lost this competition, maybe they are better off without the project. Who knows, maybe this will help dispel the taint of the arson rumours. Maybe the project would have been unwieldy anyway, full of hidden pitfalls. One of those balky, uncooperative jobs that seem to thwart every attempt to keep it on schedule. And if Jeremy has turned out to be a defector, at least this means one less salary to pay. And maybe the firm morale will improve now. Maybe he had been sabotaging them for some time. Maybe this is all for the best.

He doesn't believe a word of it.

Self-pity again, says Álvares.

They do have other projects—this is true. And if the firm is precarious, it is no more or less precarious than before. Things could be worse.

Although he can't think how.

This is not something he is good at, these attempts to be philosophical, these falsely cheerful bromides. They only make him feel glummer, more sunk into himself.

Filho de peixe sabe nadar, says his mother. The child of a fish knows how to swim.

Clearly, he isn't a fish.

Oh, go back to work, says Álvares. Design something.

Ah, yes. The one constant.

He reaches for his sketchbook.

In the next room, Matty has abandoned the television, and is prowling around on all fours, then snaking along on his belly. Once again, he is a boy on the loose, testing the forces of physics.

An ongoing physics experiment, a child of Newton. He doesn't think of himself this way—he doesn't think of himself as anything. At four years old, he is more like the apple than Newton, more like the pendulum than Foucault.

He climbs onto the sofa, and lies on his back, his head hanging over the edge. This new view of the television, upside down, is entrancing, and he lies there for a minute, kicking the cushions rhythmically with one leg. Then he slides off the sofa in stages, and rolls across the room, elbows close to his body, hands near his chin. At eye-level on the floor, he discovers a dusty cap from a pen, a rubber band, and holds on to them, still rolling.

He comes to a stop when he bumps up against the door to the balcony. Then he stands up and grasps the doorknob, lifting his legs off the floor to swing on it. Knees curled, arms tight to his chest, he hangs there, swaying back and forth. In a few seconds he is tired of this, but before he can let go, the doorknob turns far enough, and the door opens, a rush of cold air.

He lets go, surprised and pleased, and resumes his rolling, over the doorsill and onto the balcony. He lies on his back for a moment, the cement cold and gritty underneath him. He squints into the sun, and through his eyelashes, it breaks into hairs of light, refracted into faint colours. Then he rolls over on his stomach and watches a bug crawling along the cement. The ant disappears from sight, over the edge, and the boy shifts his attention to the balcony next door, through the wrought-iron bars of the railing. An elderly man comes out, stretches, and goes back in.

The boy stands up, and through the bars, catches a glimpse of something on the ground, four stories below.

A woman in a leather jacket, her hair limp, a cigarette in her mouth, is clumsily wrapping burlap around a cedar bush. He clambers onto to the wrought iron to watch her, his bare feet entwined with the bars. The cold makes him shiver a little, as he watches her head bobbing up and down as she wraps.

The woman moves out of sight to another bush, and he pulls himself up to see, hanging over the railing. Then he drops his head to feel his weight on both sides of the railing, suspended from his stomach, and rocks a little, enjoying the feeling, the heaviness in his head and arms.

But gravity has been lying in wait for him, for this moment to take its revenge. And really, how could the boy know that the ground—so far away—would be hard? In the seconds that it takes for him to overbalance, to tumble through the air, he is interested, a little surprised, his mouth a small round O, his hands floating around him, a half laugh in his throat, dizzy with the velocity of a free fall, as the balconies and windows fly past.

Then the woman is screaming, kneeling beside the motionless boy.

Chapter Fourteen

His eyes are taped closed, his face is pale as a raw potato. His small hands are open, limp. He is surrounded by medical equipment in a room that smells of rubbing alcohol. Small pads on his chest are recording his heart rate and rhythm, and an arterial line monitors his blood pressure and oxygen levels. He has an intravenous line in one arm, a cast on the other, and a black line of stitches marks one side of his head where his hair has been shaved.

A doctor comes in, a scruffy man who stands a little too closely to them, his manner a combination of earnestness and arrogance.

'Your son is still in a state of decreased consciousness, semi-comatose,' he says.

He's not our son, Eduardo thinks, but neither of them says it.

'His prognosis is uncertain.'

The doctor rubs the palms of his hands vigorously up and down his face, and then leaves, the door closing behind him with a click.

'What does that mean?' says Eduardo to the nurse who had followed the doctor in. He understood the doctor's words, but he needs someone to say it again, and again—to keep on saying it until it makes some sense.

'He's in a light coma, and we're not sure when he'll wake up,' says the nurse.

In a coma. Where is that, exactly? Maybe he is swimming in a saltwater sea of consciousness. Or maybe he is rising, instead of falling, floating on one of Geneviève's wind highways, far from *L'Hôpital de Montréal pour enfants.*

True, he is taking a long time. A good, long sleep, Eduardo thinks dully. At least he won't be tired.

A coma? says Patrick. A coma is a luminous cloud of particles that surrounds the head of a comet.

'Marie-Thérèse is on her way,' says Geneviève. She says this in the same frozen way she has said everything, although she hasn't said much.

Eduardo nods. He is speechless, his mouth dry. He feels as if a knife has sliced through a world that has suddenly become spongy—the sky, buildings, trees, even time itself have become gel-like. He sees them wrinkling for an instant as the knife comes down, and then regaining their shapes as the cut is made. Except that there is nothing on this side of the knife that he recognizes.

How could he have let this happen? The thought makes him sick with horror. He should have known—of course he knew—that children can fall, have accidents, can be injured (or killed, whispers a voice). But this means nothing, the kind of knowledge that is too unspecific, too abstract to have made an impression, at least on him, saddled with a temporary child who skimmed in and out of the margins of his existence, a child he saw out the corner of his eye. And to know that a child can fall was no preparation for a child actually falling—the knowledge almost invisible until it turned into a grotesque fact. Although even now, it seems incomprehensible, surreal.

Pourquoi cette mort? says the pamphlet. *Cherchez un sens à cette perte.*

He isn't dead, he reminds himself again. Yet.

Yet. He feels a surge of dread, the alkaline taste of it in his mouth.

How is it that the existence of this child—this collection of physical and mental processes—could rest on such a hair-trigger? That a few moments of inattention could cause this unimaginable disaster?

There must be an answer to this question. There must be some sense to this, some meaning to it.

No, says Patrick. It's just a game of chance.

But chance is too amoral, too ruthless.

True, says Patrick. Hence religion. Or law. Anything to organize chance, turn it into something explicable, or at least redistribute the consequences.

There must be some meaning to this, says Eduardo stubbornly, desperately.

Meaning is just an invention, something poured into things, says Patrick. A kind of secular faith, so we don't have to look chance in the eye.

What's wrong with inventions? says Geneviève.

Yes, says Eduardo. What's wrong with inventions? Isn't being itself a kind of act of invention?

Believe what you want, says Patrick.

But invention is different from belief, thinks Eduardo. Isn't it?

Maybe. Possibly. Perhaps. For someone like him, anyway, who measures his existence in buildings, in their structures, forms and spaces, their skeletons and skins. These are his anchors against the slipperiness of time and place, the capriciousness of temporality and geography. But he is losing this battle, the battle against things dissolving, disappearing.

He realizes suddenly—starkly—that there was never any possibility of winning it.

A desolate panic overwhelms him—he feels as if things are evaporating around him, as if he is sliding into an airless nothingness.

Fac pro viribus, says Thanatos. Act with all your strength.

Easy for you to say, says Eduardo bitterly.

What about the other possibilities? says Geneviève unexpectedly.

Like what?

Temporary footholds. Small beachheads. Like warm breath on the skin. A flight on an ironing board. A catch of music.

Footholds? Difficult to imagine—he is a doubter, his doubts are his intimates, his allies—he is not about to abandon them now. Although if footholds are even remotely possible, if there is even an iota of truth to this, then possibly—conceivably—he might consider looking for more of them. If Matty survives this. If he himself survives this. If they survive this. One thing is certain—there are no footholds on this side of the knife. On this side of the knife, there is only one terrible clarity, one terrible fact.

Wake *up*, he says silently to the small figure on the bed.

Réveille-toi, he says sternly.

Keep breathing, he says hopelessly.

Chapter Fifteen

This is a riddle, this is a trick, Geneviève says to Matty. Isn't it?

Matty doesn't reply.

I know if I stay here, perfectly still, you'll get up off that bed and start whirling around.

He says nothing.

She feels a black sob rising in her windpipe, and she forces it back down.

Why didn't she say to Eduardo: *watch him like a hawk?* Why didn't she say: *lock the doors*? *Watch the balcony*?

She is furious with him, but she also aches for him, for his obvious despair. And she can't forgive herself.

Why didn't she say: *never let him out of your sight?*

She is even angrier with him—unreasonably—for letting Matty fall in this particular way—this way that drags her into it, that entangles her, too. But it entangles her for a reason—she should have said something. Or made sure the balcony door was locked.

And her anger is hamstrung by the existing balance sheet between them, anyway. She is not entitled to be angry at him with so much guilt on her side of the ledger.

Surely not guilt? says Patrick.

Regret, self-reproach. Whatever it is.

So that's it, then? says Patrick.

That's it.

Absolutely, categorically? No possibility of appeal?

No appeal.

She knows Matty's fall isn't a punishment—what a primitive thought. But the effect is so savage, so swift that it might as well be—more than enough to shock the most equivocal mind into decision. Or to induce the most elemental form of superstition. *If I stay here, perfectly still.* This horrifying thing, this fall provides such a bald diagram of consequence that everything preceding it is implicated. All events beforehand are now potential causes, logical or not.

She has thought about making a confession of sorts, to alleviate Eduardo's burden. No, not that confession. This one: *I found him on the balcony before. I should have said something to you.*

But I didn't let him fall, she thinks. Was it only luck that I caught him in time? Or was it a question of paying attention?

Does it matter? There will be more than enough blame to go around when Marie-Thérèse arrives. Although after the first stunned reaction, she was surprisingly subdued on the telephone. This won't last long, though. Geneviève shrinks inwardly, thinking of what Marie-Thérèse will have to say. And—speaking of blame—who could blame her?

And what can she say in return? *I'm sorry. I'm so sorry. I'm so very, very sorry. I'm so profoundly, deeply sorry. I know that nothing excuses what happened, but I am so unutterably sorry.* She pictures a long, whistling string of *sorrys* swirling and twisting through the air, growing longer and longer, but still pathetically, excruciatingly inadequate.

It's my fault, says Eduardo.

It's partly my fault, says Geneviève.

It's not my fault, says Patrick. Or not that I know of.

What about the *but for* test?

Proximate cause is a tricky thing, says Patrick.

Proximity is even trickier, says Geneviève. Here we are, bouncing off each other like crazed molecules in perpetual motion.

Just like that. But it isn't just like that. She is stricken by the loss of Patrick, the loss of this chance for—for what? For this *thing*, heady, hazy, strong. But her loss is dwarfed by Matty's fall, by the prospect of his loss-to-be-perhaps, by this infinitely slow motion catastrophe.

She feels herself becoming grimmer, hardening. A necessity, something required in this substituted landscape, this imposter of a landscape in which tragedy can slip in so easily, so simply. So casually. Where a small life can disappear as easily as losing a button. This new organization of events, this new configuration of things requires something steelier.

But she looks over at Eduardo, his anguished face, and she feels swamped in inarticulate feeling for him.

He's not dead, she wants to say to him. At least that's something. *He's not dead.*

Instead she takes his hand. He looks up startled, a flicker of gratitude.

She glances back at the figure on the bed. Has anything moved? A finger? An eyelash? No.

Let me tell you a story, she says to Matty. A fungus story. You'll like this. It's about a fire-loving fungus, *geopyxis carbonaria.* Small orange cups that grow after a wildfire. Is it the change in soil? The presence of charcoal? The lack of competition from other plants? One theory is that the spores are actually stimulated by the heat of the fire.

You know what they call a fungus like this? A pioneer fungus. An early colonizer.

Picture this—the fire-scorched terrain, acres of ash, black tree skeletons. But underneath all this, the heat has woken up sleepy fungus spores, and they begin germinating away, producing fruiting bodies, flourishing in the sterile soil.

See? she says. Isn't that interesting? Isn't that amazing?

Now wake up.

Matty says nothing.

She feels as if there is a quiet scream trapped in her head.

Saint Joseph, says her mother. Patron saint of children.

Saint Francis, says Geneviève. Patron saint of flyers.

Saint Aurelius, says her mother. Patron saint against head injuries.

Saint James, says Geneviève. Patron saint of the island of Montréal, stuck in the middle of the St. Lawrence.

Ah, says her mother, taking off her print wrap, sliding a pin into her hair, and patting it. Then she is gone.

Saint Jerome, patron saint of the abandoned.

Who abandoned whom? says Saint James severely.

This is why I'm an atheist, says Geneviève.

Chapter Sixteen

A few days later, it begins to snow. The city is quickly deep-frozen, thousands of people buried in the warmth of their apartments, their windows caulked, their walls stuffed with old newspaper. Here and there inside the walls, the desiccated carcass of a long-dead mouse, an empty box of carpet tacks, a rusted tin of baking soda.

But no amount of caulking can keep out a Montréal winter, a massive, dumb white beast of a season that lumbers in and sits down heavily on the city. The city is choked with snow, every street, every alley. Cars are buried in white humps. At night, massive plows drone along the streets in ghostly convoys, pushing the snow into banks. During the day, people scale the banks, their boots patterned in salt stains, an arm held out to grasp a telephone pole for support.

Snow, they think, when they wake up.

Snow, they think, when they go to bed.

Snow is their condition. Snow is their witness. Snow is the abusive relationship they cannot shake.

And in the middle of all this snow, a small boy sneezes, a dry, fluttery sneeze. Then he tries to open his taped eyes, and starts to wail.

Eduardo is the one who sees him—hears him. They are taking shifts, but Geneviève is dozing in her chair—she has been more tired than usual lately. He bends over and lifts the boy up, wires and all, and then sits down on the bed, the boy clutched in his arms. Then he is crying, too—slow, ferrous tears that drop silently onto the boy who is now howling with indignation.

Geneviève wakes with a start. Seeing Eduardo crying, rocking the boy fiercely in his arms, for an agonizing second she thinks the boy must be dead. Then she realizes he is not only alive but the source of the howling, arms flailing as he tries to wriggle away from Eduardo. She laughs, a great, sobbing laugh, and for a moment, feels her grief spinning upward, out of her body, as if layers of calcification have been removed from her chest cavity, leaving only an enormous lightness. Then she is beside them, holding both of them, laughing and patting the boy, gently peeling the tape off his eyes.

Time is an accident connected with motion, says Patrick.

What pompous ass said that? says Eduardo.

Maimonides.

A Spaniard, snorts Eduardo.

'I have a new deadline,' she says to the department chair in January.

This is true, but it doesn't entirely explain her sudden decisiveness, her new determination. She seems to have lost her hesitancy, her vacillation—without knowing how and without caring why.

I feel less—ambivalent, I guess. More honed. You know what I mean? she says to Luc.

Barely, says Luc.

Like I could tunnel through the debris now, pushing everything out of the way.

Now I have no idea what you're talking about.

She is pulling together her research, reorganizing it to circumvent any gaps. She works on it intensely, clearly, making connections in small, bright bursts. Weak patches, false starts and dead ends—she deletes them instantly without remorse. She cuts away excess, pulls back on conjecture, tightens up analysis. Even Eduardo is impressed by this burst of activity.

She keeps on going, pouring over charts, fine-tuning algorithms and reframing the text, wrapping it around cool, hard seams of data. Finally, she sends it off to an international mycology journal.

The department chair is so startled and delighted he forgets to ask what the new deadline is.

A small crowd has gathered on a stretch of muddy earth, a few patches of snow still left among the whorls of dead weeds and grass. The April sun is weak, and Geneviève, starved for spring, takes a breath of the mild air. She is resting one of her hands on the bulge below her coat. The baby—woken from its lazy, fetal dreams—is stirring, still blind but growing, genetic spirals replicating over and over, heedless of their not quite certain origins.

Do we know whose baby it is? Patrick had said to her casually a month ago.

Yes, she said. We do.

He looked inquiringly.

It's mine, she said.

What if it looks English?

My father's genes, of course.

She said this with a kind of flinty calm that cut off any other questions.

They are wary acquaintances now—she tries to be pleasant but holds herself at a distance, determined to avoid the sticky residue of their liaison. But he is the one who spends less time with both of them now.

The new girlfriend, says Eduardo, pleased with himself for this insight.

A spurt of laughter escapes Geneviève, and then when he looks annoyed, she says hastily: I'm sure you're right.

Beside her now, Eduardo is impatient for the ceremony to start. She doesn't mind waiting—the newness of the air, even the feebleness of the sun seems to make things more distinct, more lucid. If she wanted to, she could reach out and touch the fabric of Marie-Thérèse's jacket in front of her, Marie-Thérèse who has finally—months later—if not forgiven them, then at least acknowledged the difficulties of keeping her live wire of a child safe. The child himself is jumping up and down on the spot beside her, only stopping to run circles around Luc's daughters. One of them retreats into her father's arms, while the other starts chasing Matty. Eduardo's father is behind them, holding his wife's arm. A few feet away, a cautious distance, Patrick is standing with the new girlfriend, a personal injury lawyer with a misleadingly sleepy face, earrings in the shape of tiny cockatoos, and a slow smile.

At the front of the crowd there is a large signboard with a rendering of the mausoleum. The finely cut stone rectangles are still there, but there is a glass shroud around one side. The rectangles are quieter, heavier, but they are delicately placed in relation to each other, and they form unusual angles, angles that fall into place effortlessly. The sweep of glass has been caught, mid-motion, in counterpoint to the stillness of the stone. The solemnity, the bright, cold lines have been carried over from before, but now there is something graver about them, more absolute. A finality, something tinged with bittersweet mourning, a profound incarnation of memory and regret. As if the building had caught the very second of ultimate departure, of loss, but also imbued it with a soaring transcendence, a feeling of flight as well.

A knot of people arrives, moving purposefully towards the front of the crowd. There is a rustle of expectancy, and then the cemetery director picks up a microphone, and begins mumbling something into it.

'Turn it on,' someone calls from behind them, and the director fiddles with it for a second. Then his tinny voice suddenly breaks out over the crowd.

'Is it working now?' he says in French.

'Yes, yes.'

He begins welcoming them again, clearly pleased. He makes an awkward joke about the mud, and the crowd laughs obligingly. Then he is introducing Eduardo, and there is a thin round of applause. Eduardo moves up to the front to take the microphone. He says a few words in French, little more than a graceful nod to the cemetery director, the rest of the Board, the city councillor, something about what an honour it is to work on the project. Then he hands the microphone to the city councillor and comes back down to stand beside Geneviève.

The city councillor is an articulate, weary man, using phrases prepared by someone else—*renouvellement urbain, l'édifice élégant, place de la tranquillité et la dignité.* He uses them well, though, giving them more meaning and subtlety than they deserve. A photographer to one side has heard him before, and he looks around in a desultory way.

'Urban renewal—not the words I would use to describe a mausoleum,' says Geneviève, and Eduardo laughs.

The councillor looks around uncertainly, mid-sentence, for the source of this interruption. Eduardo gestures towards him, and reassured, the councillor continues.

Now he is winding up, and he takes a shovel from an aide and digs it into the ground, stepping on it with his foot. He lifts up a shovelful of cold mud and pauses, and the photographer springs to life, taking a number of photographs in rapid succession. The crowd produces its thin round of applause again. Then the councillor jams the shovel back into the ground, and the crowd begins breaking up.

'Congratulations,' says Patrick, who has drifted over to Eduardo.

'A little premature,' says Eduardo.

'Surely getting to the ground-breaking is an achievement?'

'Yes, but now it will be one crisis, one unpredictable problem after another. They'll find something strange during the excavation, old foundations, a buried creek bed. Or some critical material will be out of stock for months. Or one of the trades will go on a rotating strike, probably the electricians. Or the concrete pours will be rained out. Or likely all of those things. And when I argue with the contractor, he'll throw up his hands, as if he had never seen the construction drawings, and say *What do you expect? It's a difficult building.*'

'What *do* you expect?' says Geneviève suddenly.

She turns to look at him, and Patrick turns his head as well.

Eduardo hesitates, surprised.

'What do I expect?' he says, finally, ruefully. 'I expect exquisite buildings, fearless buildings, buildings that ask questions, that change understanding.'

Geneviève puts her arms around him from the back, the side of her face resting on one of his shoulder blades.

'That's all?' says Patrick wryly.

'No,' says Eduardo. 'But that's enough.'

Acknowledgements

The first (and always the first) person to thank is my partner, Peter Dorfman, who is extraordinarily smart, kind, daring and funny. This is no exaggeration. He is so original, so unusual that if I wrote a character based on him, no-one would believe it. *I* wouldn't believe it. But there he is, undeniably true. Then our children, Julia Dorfman and Daniel McCormack who have each grown into such unique, warm-hearted, remarkable people that it makes me wildly happy just to think of them. I would detail all their astonishing and lovely characteristics, but it would sound like mere parental pride. Other family members include my twin sister, Naomi McCormack, a filmmaker whose talent and courage knock me flat, and my mother and late father, who were frequently ahead of their time.

But back to the book. I couldn't have written it without Barry Sampson, a brilliant architect with myriad awards to prove it, although he is not Eduardo—the only trait they share is an incorrigible passion for architecture. His firm Baird Sampson Neuert has designed several mausoleums, among many other beautiful buildings—the final mausoleum design in the book is loosely based on one. He is also a dear friend (as is his partner, fellow writer Judi Coburn), with the result that he showed me around the working life of an architect, introduced me to the buildings of Álvaro Siza, and allowed me to ply him with wine and ask questions relentlessly.

I am grateful as well to Daniel Paquin and the late Yolande Doyon for things Québécois, Vasco dos Santos and my neighbours in Little Portugal for things Portuguese, and Olga Van Kranendonk for things that have to do with quartets, all of whom contributed not only their expertise but bits and pieces of their own experiences.

People I cannot thank enough include John Metcalf, a generous, inspiring and encouraging editor who loves writing beyond reason, my excellent agent team over the last few years, Bruce Westwood (wise and gracious) and Carolyn Forde (astute and tireless), Dan Wells and others at Biblioasis for doing what they do so exceptionally well, Nino Ricci, who gave me perfectly tuned comments on an earlier version of the manuscript, Stephen Henighan, who provided French and Portuguese editing as Biblioasis' translation editor, and Christina Thompson of the Harvard Review for publishing my stories at pivotal moments.

Then there are the friends who over the years have provided moral support or who have added—in varying ways and degrees—to the strange and elusive store of ingredients that make someone a writer: Karen Andrews, Susan Ballantyne, Patsy Berton, Donna Bobier, Laura Bradbury, Don Chiasson, Heather Chetwynd, Marion Cohen, Paul Copeland, Barbara Dresner, Wayne Ellwood, Shelley Gavigan, Rico Gerussi, Sonja Greckol, Bram Herlich, Kris Heshka, Shin Imai, Barbara Jackman, Mary Kainer, Michael Kainer, Bob Kellermann, Murray Klippenstein, Joy Klopp, Brent Knazan, Kathy Laird, Ron Lebi, Sherry Liang, Liz Martin, Diana Meredith, Marilyn Nairn, Kathleen O'Neil, Ruth Shamai, Roman Stoykewych, Susan Stewart, and Liz Woods.

And my friends/colleagues at the Faculty of Law, who admittedly did not have much to do with the book, but whose camaraderie, working chemistry—and in some cases, dark humour—allowed me to go home at the end of the day with a few brain cells left for writing: Whittney Ambeault, Karen Bellinger, Jutta Brunnée, Lisa Cirillo, Aleatha Cox, Hilary Evans Cameron, Chantelle Courtney, Sara Faherty, Tracey Gameiro, Nikki Gershbain, Claire Hepburn, Kate Hilton, Ed Iacobucci, Ivana Kadic, Jane Kidner, Judith Lavin, Natasha Mackley, Renu Mandhane, Kristen Marshall, Cheryl Milne, Mayo Moran, Emily Orchand, Mariana Mota Prado, Sarah Pole, Andrea Russell, Promise Holmes Skinner, Kim Snell, Archana Sridhar, Michael Trebilcock and many others.

The market in the book borrows characteristics from a number of real markets, including Atwater and Jean-Talon in Montréal and markets in Málaga, Barcelona, Lisbon, Amsterdam, Paris, Apt, Istanbul, Palermo, Siracusa and elsewhere; the Montréal Botanical Gardens really do include a Pavilion of Infinite Pleasantness, a Tower of Condensing Clouds, a Garden of Weedlessness and an adjacent Insectarium which gave rise to their respective descriptions, but not as far as I know, a fungi room; needless to say, Geneviève's biology department and faculty members are entirely fictional; the cemeteries in the book have been influenced by several cemeteries, including Notre-Dame-des-Neiges Cemetery, Mount Royal Cemetery and Montparnasse Cemetery (and the Cemetery of Pleasures in Lisbon really exists); the cello trio who played on mountain tops and cathedrals is Extreme Cello; the legend about *pwdre ser* comes from the Edinburgh Geological Society and the foul jelly quote is from *The Prose Works of Sir Walter Scott, Volume 4* (Edinburgh: Robert Cadel, 1834); the vegetable love quote is from 'To His Coy Mistress' by Andrew Marvell; some of the information about Gordon Matta-Clark came from the Canadian Centre for

Architecture, which is also the inspiration for Eduardo's lecture and the source of the fact about Montréal's cultural divide in building materials; mycological information came from a variety of sources, including the Mycological Society of Toronto, the North American Mycological Association, ScienceNews, Botanical Electronic News, Kew Royal Botanical Gardens, the Harvard Gazette, Waldwissen.net, Treehugger.com, and the Virtual Museum of Canada; Portuguese proverbs came from my neighbours, Quemdisse.com, Walter K. Kelly, *Proverbs of All Nations* (London: W. Kent & Co., 1859), Dicionário inFormal.com and Duolingo.com; and the driving-in-a-snowstorm explanation of the radiant point is from several websites—including, for example, Cloudbait Observatory in Colorado. Of course, all errors are mine.

About the Author

Judith McCormack was born outside Chicago, and grew up in Toronto, with brief stints in Montréal and Vancouver. She has several law degrees which have mostly served to convince her that law is a branch of fiction. Her first short story was nominated for the Journey Prize, and the next three were selected for the *Coming Attractions* anthology. Her collection of short stories, *The Rule of Last Clear Chance*, was nominated for both the Commonwealth Writers Prize and the Rogers Writers' Trust Fiction Award, and named one of the best books of the year by the *Globe and Mail*. Her work has been published in the *Harvard Review*, *Descant* and *The Fiddlehead*, and one of her stories has been made into a film by her twin sister, Naomi McCormack, an award-winning filmmaker. Her most recent short story was anthologized in *14: Best Canadian Short Stories* and was recorded in a spoken word version by *The Drum*.